SCIENCE
EDUCATION
PARTNERSHIPS

SCIENCE EDUCATION PARTNERSHIPS

MANUAL FOR SCIENTISTS AND K-12 TEACHERS

EDITED BY ART SUSSMAN, PH.D.

PREFACE BY BRUCE M. ALBERTS, PH.D.
PRESIDENT, NATIONAL ACADEMY OF SCIENCES

PUBLISHED BY
UNIVERSITY OF CALIFORNIA, SAN FRANCISCO

SCIENCE EDUCATION PARTNERSHIPS:
Manual for Scientists and K-12 Teachers

Edited by Art Sussman, Ph.D.

Copyright © 1993 by

University of California, San Francisco
3rd Avenue and Parnassus Street
San Francisco, California 94143

Library of Congress Catalog Card Number 93-084434

ISBN 0-9635683-0-2 Cloth
0-9635683-1-0 Paper Trade

Manufactured in the United States of America

Design by Community Graphics/San Francisco Study Center

CONTENTS

STUDENT CENTERED ACTIVITIES

ACKNOWLEDGEMENTS

The writing and publication of this book was made possible only through the efforts of many dedicated and generous people and organizations. In particular, the American Honda Foundation and its Manager, Kathy Carey, provided vital financial assistance for the formation and growth of the UCSF Science and Health Education Partnership as well as for the initial publication of this book.

None of us anticipated that our original commitment to produce a "how-to" science education partnership "booklet" would grow into a book of this size and scope. As we became more involved in the world of partnership activities, we became aware of the many different kinds of programs and resources that universities, science centers and businesses share with K-12 schools, teachers and students. Fortunately, many wonderful people were willing to take extra time in their all-too-busy lives to write about their programs and activities. They shared with us the conviction that more people need to become aware of and learn from our collective efforts.

Finally, I would like to acknowledge the assistance of many teachers and education administrators who inspire us and their students by their expertise and personal commitments to work far beyond the call of duty under extremely difficult conditions. Collaborating with them and seeing the positive results with their students are the greatest rewards of our partnership efforts.

The articles in this book reflect only the views of their respective authors. The American Honda Foundation, the University of California, the Science and Health Education Partnership, and the institutions that employ the authors are not responsible for and do not necessarily agree with the contents of these articles.

Art Sussman
San Francisco, California
July 1993

PREFACE

WHY SCIENCE EDUCATION PARTNERSHIPS?

BY BRUCE M. ALBERTS

Science education is special in several ways. First, science is presently greatly undervalued in our schools, generally being treated as an optional activity, rather than as a core subject that should occupy 20%-25% of total class time, starting in kindergarten and continuing through 12th grade. Second, the type of science education we want emphasizes hands-on problem-solving that actively engages students, and it requires special materials that must be organized and maintained. Third, precollege science education is often overseen by school district personnel who are not experts in precollege science education themselves, and who are unlikely to be effective advocates for the new resources that will be needed to create excellence. For all of the above reasons, we badly need the active participation and interest of informed scientists and engineers in today's schools.

One of the most notable experiments in curriculum reform was Man: A Course of Study (MACOS), an innovative social science program designed in the 1960s for elementary schools. The story of its rise and fall is told in a fascinating recent book written by Peter Dow, the MACOS project director. In assessing its failure, he concludes: "The way we usually market educational innovation leaves little room for enlisting the power of teachers or providing for a continuing relationship between schools and universities. Teachers are typically confined to their classrooms with schedules that preclude significant interaction with other educational professionals. When this partnership disintegrated or ran into resistance from the authority structure of the school, much of the power of the course to bring about positive change in the classroom began to disappear.... The decline of the scholar-teacher partnership was one of the major reasons for the demise of the science curriculum reform movement." (Peter B. Dow, *Schoolhouse Politics: Lessons from the Sputnik Era*, p. 257, Harvard University Press, 1991.)

"Effective partnerships are the only hope for lasting systemic change in precollege science education."

One can restate the above conclusion in a way that will sound familiar to all students of cell biology. If left alone, schools, like chemical systems, will tend to drift toward their lowest energy state. The current school culture tends to be dominated by overworked teachers, uninteresting but easy to use textbooks, and school rankings by standardized exams. In this environment, it is not surprising that science courses have typically become courses in science vocabulary that interest nobody and make science seem like drudgery — even to science junkies like myself. Avoiding this situation, so as to maintain excellence in our schools, will require a continuous input of energy.

I have learned a great deal, both about education and about partnerships, in the past five years. When the Science and Health Education Partnership at UCSF began in 1987, I viewed it largely as an effort of an institution that is rich in science, equipment, and supplies to help our needy colleagues — the science teachers in the San Francisco public schools. I had no idea how much the scientists would learn from these teachers about effective teaching methods, nor did I anticipate the invigorating effect on the university of this form of public service. But most of all, I underestimated the potential importance of the partnership for the schools. My hope had been the very modest one, in retrospect, of making perhaps a 5% difference in the science education offered in the San Francisco schools. For the reasons expressed by Peter Dow, I now view effective partnerships between scientists and precollege science teachers in a completely different light—as the only hope for lasting systemic change in precollege science education and, therefore, as an important national priority for the United States.

In our country, there is a strong tradition of local control of our public schools. The resources available for public education are limited, with many competing demands. Good science teaching will always be rela-

---

2

tively expensive, and each school district will need knowledgeable and persistent science advocates if it is to maintain an emphasis on high-quality science education. Who else but local scientists and engineers can be counted on to be effective in this role? Of course, not just any scientist or engineer will do. To be effective, these volunteers will need to be sensitive to the environment of the schools, closely tied to some of the best science teachers in the school district so that they can be advocates for their needs, and able to serve as connectors to important resources that the schools cannot otherwise obtain. They should also be well informed about models of quality science education elsewhere that can be transferred to their local schools.

The National Academy of Sciences has begun an effort to educate all interested scientists about the opportunities and challenges of precollege science education through workshops given at the annual meetings of the many different scientific societies (the microbiologists, biochemists, cell biologists, chemists, neurobiologists, and so on). The aim is to create a national network of informed volunteers who act locally to improve science education in their public schools. These individuals will be much more effective in causing change if they function as part of a local partnership, rather than representing only themselves. In my view, we need to increase the number of such partnerships by several orders of magnitude in the next decade, so as to involve most of the science-rich institutions in the United States (whether college, university, local industry, or museum). Of equal importance, like UCSF's partnership, each partnership must become more ambitious. Rather than being satisfied with providing supplementary help to local teachers, the major partnership goal should be to provide long-term support for the systemic changes that are needed to create excellent science education in the local schools.

A great deal of patience will be needed, inasmuch as any major change is likely to require on the order of 10 years—much longer than will seem reasonable to any scientist who has not previously attempted to have a major impact on a large conservative institution, such as a school district or a university.

So let us begin!

A high school student teaches a lesson on the elements to seventh graders in UCSF's Science and Health Lesson Plan Contest (photo courtesy of Margaret Clark).

(Right) An industry mentor assists a teacher using a specialized microscope as part of his summer employment in Raytheon's Semiconductor Division (photo courtesy of IISME).

INTRODUCTION

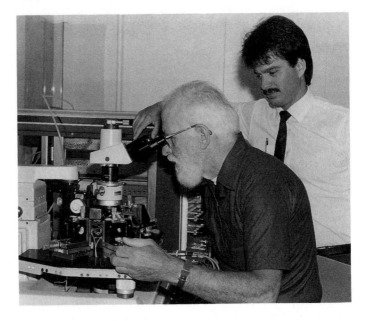

ABOUT THE AUTHORS

BRUCE M. ALBERTS, PH.D. ◆ As chairman of the University of California at San Francisco (UCSF) Department of Biochemistry and Biophysics, Bruce initiated the Science and Health Education Partnership. He is currently president of the National Academy of Sciences.

WILLIAM ALLSTETTER ◆ William is the editor of *Today's Science on File*, a science magazine aimed at junior and senior high school students.

GREGORY ARENT ◆ Greg is a medical student in the UCSF School of Medicine. He volunteered extensively in the MedTeach Program of the Science and Health Education Partnership.

JACQUELINE BARBER ◆ Jacquey is director of the Great Explorations in Math and Science (GEMS) program at the Lawrence Hall of Science. GEMS is famous for its many curriculum modules that help teachers do hands-on science in a wide variety of topics.

KATHARINE BARRETT, PH.D. ◆ Katharine is Director of Biology Education at the Lawrence Hall of Science.

SUSAN BRADY, PH.D. ◆ A botanist by training, Sue coordinates a variety of education outreach programs at the Lawrence Hall of Science.

KATHERINE CAREY ◆ Kathy is the manager of the American Honda Foundation.

PAM CASTORI ◆ Pam is co-director of the California Science Project. A high school science teacher, she left the classroom to become a science specialist in the School of Education at the University of California, Davis.

MARGARET CLARK, PH.D. ◆ A physical chemist by training, Margaret is currently director of the UCSF Science and Health Education Partnership.

JEAN COLVIN, PH.D. ◆ A molecular biologist by training, Jean is director of the University Research Expeditions Program housed at the University of California, Berkeley.

LANE CONN ◆ Lane is director of the Teacher Education in Biology Program. He has conducted research in cell biology, biochemistry and physical chemistry, and has taught both biology and physical sciences in the California public schools.

GLENN CROSBY, PH.D. ◆ Since 1967, Glenn has been at Washington State University, serving as professor of Chemistry and Chemical Physics. He is a research scientist who has been extensively involved in teacher education programs.

RAMON CORTINES ◆ As superintendent of San Francisco Unified School District, Ray promoted many partnerships between the schools and outside organizations. He is currently consultant to the U.S. Secretary of Education.

CHRIS DELATOUR, PH.D. ◆ A physicist by training, Chris helped establish and direct the Teacher Institute of the Exploratorium in San Francisco.

MARIE EARL ◆ Marie held a variety of positions in the education and science policy arena with the American Electronics Association, Stanford University, and the U.S. Senate. She currently serves as executive director of IISME, Industry Initiatives for Science and Math Education.

STEFAN GAIR ◆ Stefan is the curriculum coordinator for Math and Science at the Pittsburg (California) Unified School District. Previously, he was an elementary school principal for five years.

SHELBY GIVENS, PH.D. ◆ Shelby is director of the Northern California Regional Center of the Mathematics Engineering and Science Achievement (MESA) Program.

DENNIS HARTZELL ◆ Dennis is director of Corporate and Foundation Relations in the UCSF Development Office. He has been very instrumental in raising funds for the Science and Health Education Partnership.

BRIAN KEARNEY, PH.D. ◆ A molecular biologist by training, Brian is director of Education and Research for Industry Initiatives for Science and Math Education at the Lawrence Hall of Science.

KAREN MENDELOW ◆ A biologist by training, Karen has been involved in education and teaching outreach programs in science museums. She has worked with the Exploratorium Teacher Institute since 1987 as program manager, instructor and librarian.

MIKE MORGAN, PH.D. ◆ Mike, a postdoctoral neuroscientist, has volunteered extensively in many different activities in the UCSF Science and Health Education Partnership.

LEON LEDERMAN, PH.D. ◆ Leon received the 1988 Nobel Prize in physics. Elected in 1991 as president of the American Association for the Advancement of Science (AAAS), he devotes considerable effort to improving public understanding of science.

JOANNE MILLER, M.A. ◆ JoAnne was health director of the UCSF Science and Health Education Partnership. She also served as a member and president of the Board of Education of San Francisco Unified School District.

DEAN MULLER ◆ Dean is in the Science Department at Mark Twain Continuation School, a high school in San Francisco.

KARIN ROSMAN ◆ Karin has held a variety of positions at Lawrence Hall of Science, coordinating science education outreach programs for teachers. She is currently education coordinator for IISME.

MARK ST. JOHN, PH.D. ◆ A physicist by training, Mark has become deeply involved in precollege science education programs. He is founder and president of Inverness Research Associates, a company that specializes in program evaluation.

ZINA SEGRE ◆ Zina is a physiologist who has held a variety of science-related positions in the Bay Area as a teacher and a medical writer. She currently manages sponsor relations for the IISME Fellowship Program.

KATHRYN SLOANE ◆ Kathryn is assistant research educator at the Graduate School of Education at the University of California at Berkeley. She specializes in evaluation, qualitative methodology, and the assessment of home and classroom learning environments.

CARY SNEIDER, PH.D. ◆ Cary received his B.S. in astronomy and his M.S. and Ph.D. in science education. He is currently director of Astronomy and Physics at U.C.'s Lawrence Hall of Science, where he develops curriculum materials and conducts workshops and institutes for teachers.

SARAH SPENCE ◆ Sarah is a medical student in the UCSF School of Medicine. She volunteered extensively in the MedTeach program of the Science and Health Education Partnership.

DAVID STATES ◆ As an academic coordinator at UCSF, David worked for three years as associate director of the Science and Health Education Partnership. He is currently in Germany with his wife who holds a postdoctoral position at the University of Heidelberg.

ART SUSSMAN, PH.D. ◆ Trained as a biochemist, Art was science director of the UCSF Science and Health Education Partnership. He currently is director of the Far West Regional Consortium for Science and Mathematics Education serving Arizona, California, Nevada and Utah.

HECTOR TIMOURIAN, PH.D. ◆ Hector has 10 years experience in creating and implementing education outreach programs at Lawrence Livermore National Laboratory.

A FRAMEWORK FOR SCIENCE EDUCATION PARTNERSHIPS

BY ART SUSSMAN

An early criticism of the partnership movement was that it involved a lot of people running around doing things, but that the whole added up to less than the sum of its parts. Today, on the other hand, one can make a convincing case that a consensus has emerged regarding the kinds of reforms that are needed in precollege science education. As a result, partnership activities tend to address the same constellation of issues in similar ways. While we have not yet reached a goal of a coordinated, coherent reform movement that maximizes the integration of partnerships with a variety of other resources, the agreements about a general framework help make science education partnership efforts more fruitful and more focused.

This framework has several key features. Perhaps most important, it aims to reach all students. Science is not a subject that is reserved for an elite group of white males who hate sports. Everybody can enjoy and succeed in science. All members of our society need to be scientifically literate for their own well-being and for the health of our society. From a society perspective, this argument is usually made in narrow economic terms, that the new global marketplace requires a technically literate workforce. More generally and importantly, our democracy depends on the ability of citizens to think well, solve problems, address technical issues, and make wise choices for themselves, our country and the planet.

The need to reach all citizens is particularly challenging with respect to groups that are underrepresented in science. Current gross inequities

---

with respect to ethnic minorities and females reflect deficiencies in our educational system and in our society. The new pedagogy must incorporate approaches and practices that encourage, rather than discourage, those who come from groups that are currently underrepresented in scientific fields. For example, what aspects of our curricula, school culture, and testing procedures are brick walls that exist simply because currently successful scientists and teachers surmounted them in the past? We need the wisdom to see these brick walls for what they are, and then replace them with guided gateways. Even more daunting is the need to overcome the devastating handicaps of poverty and disintegrating communities.

Successfully educating all students also requires a radical change in the roles of the student and the teacher. The term constructivism is often used to describe the new approach. The student constructs her or his own meaning as she or he integrates new learning experiences with prior conceptions. The teacher does not lecture and thereby pour facts and information into empty vessels. Rather the teacher acts as a facilitator who provides an environment where students directly experience rich situations and then, often working in cooperative groups, they utilize higher-order thinking and problem-solving skills to make sense of the world. The problems that they study relate directly to their lives and reflect the real world of science. Excellence and equity go hand in hand. Many of the changes that make it easier for underrepresented groups to succeed — such as replacing lessons that emphasize lecture and vocabulary with student-centered learning involving a high proportion of hands-on experiences — also improve everyone's science education.

Radical reform in science education also necessitates radical change in the nature of schools and the teaching profession. A teacher is probably one of the only professionals in America who does not have her or his own phone at work. As a profession, teachers have low status in the community, are poorly trained to do the job that the modern world requires, lack essential equipment and supplies, lack planning and preparation time, do not have the power to make the changes that are needed in their workplace, and are isolated from each other and from the adult world.

Many of the programs described in this book are models of how partnerships can address these issues. Curriculum experts at the Lawrence Hall of Science tell how they work with teachers and school district administrators to develop curricula that engage students in meaningful hands-on activities and then integrate these lessons within a multiyear course sequence. University scientists from Chicago, San Francisco and Washington state describe workshops that integrate new curricula with teacher training and with other efforts to change the existing school culture. Famous scientists regard and treat teachers as professional equals. Many programs bring the richness of the scientific world directly into the classroom for students to experience scientists as normal human beings and scientific issues as being relevant to them and within their grasp. Female and ethnic minority role models from universities, health professions, and industry prove that the technical world can be open to all.

An important feature of science education partnerships is that they are beginning to effectively weave together the different strands of the science reform movement. Earlier reform efforts tended to emphasize just one aspect of the problem. World famous experts wrote new curricula in the expectation that improved textbooks and lesson plans would fix science education. Instead the products sat on the shelves. A current popular misconception is that the tests are the main problem, and if we could replace multiple choice exams with performance assessment we would fix science education. The better partnerships recognize the many dimensions of the current situation that need to be addressed simultaneously. Teachers need instruction in new content and skills; they need lessons and materials that embody the new philosophy; they need samples of and training in alternative assessment tools; and they need support back in the school and the community or else the vast inertia of the existing school culture will doom all our efforts.

"Systemic reform" or "systemic change" is the current buzz word for this kind of integrated, comprehensive and radical change. In addition to projects at the local level, the systemic change movement has a national and a state policy component. One articulation of the systemic reform model emphasizes the description of a vision of excellence in the form of standards. Three panels under the aegis of the National Research Council (the principal operating agency of the National

Academy of Sciences and the National Academy of Engineering) are currently developing national science education standards in the areas of curriculum, assessment and teaching.

The science curriculum standards will define the kinds of experiences that produce the science learning outcomes we desire. They will also provide a broad view of what students should learn — the scientific information (facts, concepts and big ideas) as well as the modes of reasoning and proficiency in conducting science investigations. The curriculum standards will also call for students to have the skills and desire to apply this knowledge outside the classroom.

Assessment is the tool that is supposed to inform us if we are going in the right direction, if we are achieving the outcomes that we desire. Assessment plays important roles at the daily classroom level by teachers as well as at the school, state and national levels. The most common headlines about schools appear right after the announcement of the latest round of mediocre to terrible student test scores on state, national or international examinations.

In addition to being a diagnostic tool, assessment also very directly tells students, teachers and parents what the education system thinks is important. Teachers and students often make the test a focus of their efforts. If we want the entire educational community to place a high value on hands-on experiences and higher order thinking skills, we sabotage our efforts if we then grade students using multiple choice tests that reward them for memorizing vocabulary and isolated facts ("the mitochondrion is the part of the cell that makes energy"). The new science assessment standards will define protocols for assessing students' accomplishments in ways that are valid and that reinforce the learning outcomes that we desire.

The teacher is the key to effective science reform. All the other pieces can be in place, but the reform movement will fail if classroom instructors do not have satisfactory training, knowledge and support systems. Science teaching standards will define the skills and knowledge that teachers will need as well as the necessary components of professional training programs for certifying beginning teachers and providing essential staff development for the large pool of existing teachers. The new teaching standards will also define the support systems and resources that need to be in place at the school site itself.

How can these national standards help make changes at the local school sites? On a grass-roots level, they provide powerful support that reform advocates can use with school boards and education administrators at state, county and site levels. On a more formalized state level, most states adopt science curriculum frameworks that describe the parameters for science education in that state. These documents vary in their

---

Partnerships can play a vital role in translating these standards, frameworks and assessments from the realm of abstract principles to the world of actual classroom practice. As testified by the many articles in this book, science education partnerships are a very flexible tool for bringing rich scientific resources into the hands and minds of teachers and students.

---

scope and authority, but they can have great influence on classroom teaching. Many states are already revising their science curriculum frameworks so that they embody reformed standards such as those emerging from the National Research Council or are contained within the Project 2061 documents of the American Association for the Advancement of Science. Many states are also changing the tests that they administer, moving away from recall-based multiple choice examinations to more authentic instruments that emphasize higher order thinking skills and the performance of tasks that require the application of scientific skills and knowledge to solve problems. A prominent feature of the systemic reform movement is this embodiment of world class standards within new curriculum frameworks coupled with assessment instruments that are based upon and reinforce the new standards.

Partnerships can play a vital role in translating these standards, frameworks and assessments from the realm of abstract principles to the world of actual classroom practice. As testified by the many articles in this book, science education partnerships are a very flexible tool for bringing rich scientific resources into the hands and minds of teachers and students. In addition, the scientists who participate in these programs can then contribute much more meaningfully to the continuing

---

efforts to define and implement a truly effective science education system. The committees developing standards, frameworks and assessment include scientists who have some real knowledge of elementary and secondary schools because they have actually worked inside classrooms and side-by-side with teachers as a result of partnership programs.

Science education partnerships are providing the training ground for t-RNA people, a new breed of hybrid professionals who have experience, respect, knowledge and skills in both the scientific research world and in the precollege education community. Living systems operate with two languages — the nucleic acid code that is the basis for information storage (heredity) and the protein assemblies of amino acids that provide the foundation for all cellular structures and functions. Transfer RNA (t-RNA) is the translator molecule in the cell that speaks both languages and enables the information that is stored in the nucleotides of DNA to be translated into the sequence of amino acids that make up proteins. Most scientists had very atypical experiences in their schooling and live in a very different world from elementary and high school classrooms. Most teachers, especially at the elementary level, have a very limited background in science. Even those who have taken more than one college course that includes a "laboratory" section have never experienced science as it is practiced in modern laboratories. More t-RNA people are needed to speak both languages and make the vital connections across both worlds.

Many of this book's authors are these hybrid professionals who, despite institutional resistance and the frustrations of obtaining funding, have persisted in creating living partnerships. There is no single formula for creating an effective partnership. Many different models have worked in different communities. The following article by Bruce Alberts describes one general recipe for getting started.

10-STEP RECIPE FOR STARTING A PARTNERSHIP PROGRAM

BY BRUCE M. ALBERTS

The UCSF Science and Health Education Partnership (SEP) was started by myself and David Ramsay, the vice chancellor for academic affairs. I had accidentally discovered David's interest in the schools when he and I were seated together during one of those interminable, official university dinners, which I had been obligated to attend as chairman of the Department of Biochemistry and Biophysics. The fact that David and I held positions of authority at UCSF was a great help in getting the UCSF program started quickly. The individual who attempts to start such a program will generally lack the advantages that the two of us had, and I have written this "how to" essay with this in mind. If you are energetic, tenacious, and believe that the reinvigoration of public education is of critical importance for the future of our democracy, you might find satisfaction by exploiting the following plan:

1. **Find six or more outstanding teachers in a local school district.** Nothing can be done in a partnership program without first recruiting a small team of dedicated, talented teachers to provide informed leadership. You and your colleagues will be both educated and inspired by your interactions with these energetic individuals from the "front lines." Ask the first outstanding teacher you find for the names of others.

2. **Get the superintendent of the teachers' school district to support the formation of your new partnership.** Use the advice of your teacher group concerning how to proceed. Sell this idea by cit-

ing the success of partnerships elsewhere. Get the superintendent's office to provide you with the names and school addresses of all the district's science teachers. Home addresses and phone numbers are particularly useful.

3. **Organize a meeting between your outstanding teachers and a few department chairs or other recognized leaders in your organization.** Have an agenda so that something significant is decided. Hold the meeting in a nice room and provide some food. Teachers are not used to being treated well and will appreciate small courtesies! If you have found the right teachers, they will inspire the leaders of your organization and enlist their commitment for you. Use these leaders as the core of an "executive committee" that enlists your organization's support for the program.

4. **Find the resources to support the salary of a half-time partnership coordinator to organize the program and raise funds for the partnership.** The person chosen should be energetic, organized, able to write well, and must care about improving the schools. He or she will need a small budget for food, mailing, etc. The executive committee described above will find the necessary funds. You can point out the long-run benefits to your organization's local reputation and contrast the small cost of the half-time position with the much greater resources that are spent on public relations. But do not let the PR people get anywhere near the program for several years, until you have something that is really worth advertising to the community.

5. **Start your partnership program with a well-advertised event.** Have the leadership teachers meet with a few interested scientists as a teacher-dominated "steering committee" to plan an after-school event. The partnership coordinator will be responsible for mailing invitations to all the science teachers in the school district, for advertising the event throughout your organization with mailings and posters, and for arranging the room and the food. Our first event was a "mixer" during which we divided into small groups to introduce each other and to develop suggested program elements that were reported back to the entire assembly. This mixer resulted in a program focus on forming and nurturing one-on-one partnerships

between teachers and scientists. Teachers and volunteers were invited to write down what they personally might seek in a partner. Teachers and scientists who had formed preliminary partnerships registered with the program coordinator.

6. **Establish and nurture at least one core activity.** One-on-one partnerships formed SEP's initial core activity. This focus helped us bridge the abyss separating practicing scientists from precollege science teachers. The partnership coordinator helped find scientist partners for all those teachers desiring one, encouraged the volunteers to visit his or her teacher's classroom within the first month of their pairing, and found new matches for those teachers or scientists whose partners turned out to be inappropriate or inaccessible. The partnership coordinator also made a list of available sources for dry ice, petri dishes, flies, etc. for each scientist partner to use, as well as organized the donation of surplus equipment for the classrooms of interested teachers. This book describes a variety of activities that can form your initial program core. The one-on-one partnerships fulfilled this role very well for SEP.

7. **Write grants to obtain more funding.** The operation of a substantial partnership program will require resources for program expansion, including the support of several salaries for full-time personnel. With the aid of the partnership coordinator, and the support of the development office (if such exists), funds should be sought from a variety of sources interested in public education, emphasizing local philanthropists and industries. Several sources will be necessary, since these first grants are likely to average $5,000 to $40,000 each.

8. **Use meetings of the teacher-dominated steering committee to plan new activities.** Elsewhere in this book are listed many possible activities. Teachers are overworked, and it is crucial that everything done by the partnership be at their suggestion and have their full support. We have had particular success with our annual student lesson plan contest, which generates a tremendous amount of activity in the schools for a relatively small amount of prize money. It also gets many UCSF scientists into the schools for the first time each year as contest judges.

9. **Encourage the teachers to work within the school district for**

systemic change. By themselves, teachers often feel isolated and powerless to effect needed changes. The partnership, by bringing teachers together and treating them as professionals, should have the long-term effect of giving them the confidence to work together to push for more support for science in their schools. Reaching this stage will take several years. Ideally, it will encourage the school district to organize a leadership team of outstanding science teachers, which can be relied upon to make consistent, intelligent decisions about science curricula in behalf of the school district.

10. **Work with the district leadership teachers and administrators to obtain major funding to meet important district science education needs.** Collaboratively identify major obstacles to bringing about systemic change in science education. Identify solutions that the partnership can help develop and implement to overcome these obstacles, and target funding sources that are appropriate. One important area is the need, particularly at the elementary and middle school levels, to help teachers become more science literate and better trained in leading exploratory, hands-on lessons in their classrooms. Programs that address this need usually take the form of multiyear staff development summer workshops complemented by support in the classrooms during the academic year. Multiyear grants to support these activities are generally only available to school districts through partnerships with science-rich institutions, such as universities and museums. Other articles in this book describe funding sources and strategies.

PERSPECTIVE OF A SCHOOL DISTRICT SUPERINTENDENT

INTERVIEW WITH RAMON CORTINES,
SAN FRANCISCO UNIFIED SCHOOL DISTRICT

Ramon Cortinez: I tell people about this program all the time. When I talk to them about personnel, they just can't believe the kind of commitment and the professional stature of the volunteer scientists and doctors. I saw one of your people working at a middle school, helping the teacher for a little while. And I have to tell you that that teacher told me that if it had not been for you people that she would have left teaching. She is a second-year teacher, and she's damn good. And she said that it is the single concepts that the UC staff has taught her — to focus on the most important concepts rather than teach a lot of disconnected facts. She's learned how to do hands-on and not be afraid of it — and understand that noise with kids doesn't mean that the class is out of control.

This partnership is a model. And it does not have to be just in the area of science. It says what a university can do, if there is a commitment, if there is direction, and if there is leadership. And certainly this has been a rocky love affair, and you know it. It has been a love affair of a lot of lovers, to improve education in San Francisco. It's one of the things that I wanted for San Francisco, because I thought that the science program was so wanting, and now I think we are moving in a good direction.

I think one of the things that spurred science at the middle school level was the recognition — recognition of teachers, recognition of students — and that you made teaching into a thinking process and an exploration process. You made it competitive. You set it on a different kind of level. If I can use the example, you set it on the level with sports: by the activities that were created, and then the way they were recognized and rewarded.

I think that the way that you have done the elementary in-service has been unbelievable. I have seen what you did because I attended the in-services this summer. And as you know, I was so impressed, I went and got a television station to come, because I wanted them to see this kind of partnership. Now I have visited schools where teachers are using what they learned at the university. The keeping of the log that was part of the professional development this summer. I have visited schools and seen that those teachers understand the importance of keeping a log and have transferred that — they are having the kids do it.

With the professional development, I didn't feel, our people have not felt, that you were on this level up above and we were down below. I think that there has been an even playing field from the beginning. See, I don't think that you can have a good collaboration with "top-down" or "bottom-up." I think that that's where we make the mistake in restructuring in schools across the nation. When we talk about coming from above or below. I think you have to move it to a level playing field. And, there's a constant professional tension on a level playing field. It's almost as if I took a rubber band, tied it in the middle and then just pulled it horizontally from both ends. You're never apart, sometimes you're further apart, many times you are very close. But you're always connected and it's level.

Art Sussman: Some collaboratives seem to have a more institutionalized mechanism for the university and the school district people to be meeting and planning for the future. We have not done that as much.

RC: I think that often there's a lot of talking in meetings. There may not be a lot of doing. And, this one is a lot of doing. We started informally. I don't really think it originated from me or the university, it sort of grew together. And right away we started doing things — the partnerships between teachers and scientists, the student contest, the equipment distribution.

AS: Let's get into that tricky area that you mentioned — the rocky part of the relationship.

RC: I don't think you can avoid some of that. I think the rubber band has really gotten stretched at times. But, I don't think it ever completely broke, or was disconnected. I really monitored it very closely. This is one of the things I'm most proud of in my six years of this relationship. I so believe in collaboratives. The University of California was not diddling with us — this is systemic, in-depth professional development ultimately resulting in change in how science is delivered in the schools.

AS: If a program aims just to provide some services to an existing structure but not try to change the structure, then I think there's less potential for the kind of friction that we have had.

RC: I think that when you undertake a situation like this that you deal with personalities. See, I don't think the university or the school system caused difficulties. The university deals with thousands of employees, and I deal with thousands of employees. There are going to be issues and problems. I expect it. Maybe it's because I'm old. I just know it's going to happen. But I know that we'll get through it. I think that people want to own things. They want it to be their program rather than the students' program. They just can't get themselves out of the way. And that happens on both sides. You have to expect the growing pains. You have to expect personalities. Scientists and doctors are primadonas. So are school teachers and superintendents.

Maybe I did not take enough time to explain the relationship to people. That's a fault of mine. You can spend your time bringing people together, getting them ready. And I tend to cut short that step. Time is too precious for me. I'd rather have it rocky a little and get there. And the 35 years have sort of proven that you always get there. And you get there sooner by that relationship than you do sitting down and "contemplating your navel" for hours and hours. So, that's a style of mine, and others (and I agree with them) would say that's a fault of mine. As the marriage counselor, I was never going to let the marriage break up.

AS: How do you get more people involved, not just the enthusiastic teachers and principals?

RC: Let's talk about the elementary schools. I checked how many went to the first in-service. Fifty-two schools were represented. I can tell you now that all 72 elementary schools are represented. They may not all be there because they want to be there, but I said that there would be representation from September of all 72, or the principals themselves would attend. Now see, you can't always have it all volunteers, sometimes there needs to be a person with a big stick for the right reason, or a benevolent dictator. Now I'm sure that many of those 72 schools are at various levels but they're all there. See, I don't believe in projects. Let me clarify that. If the program is good for all third-graders, then it's good for all third-graders, and shouldn't be just for these kids over here. Because generally the kids that need it the most, are not the ones that are going to get it.

That's the reason that I didn't suggest to the middle schools, "Well if you want to have three years of science, you can." And when somebody over here east of Twin Peaks said well our kids don't need it, I said, "Bull." I said we will have science all three years for all of the schools. Now there are schools in this district that have made greater advances in the three years: in the content, in the curriculum, in the materials and in the professional development — all of that. But all of them, all of the kids are getting a little. I'm depending on the kind of in-service, the kind of plateau that I believe that we are in this collaboration, that we will all get to a satisfactory level at some time. I don't believe because we have a good principal over here and some good teachers here that we just do those. It's not the same quality at every school, but it's there.

AS: So it sounds like you're saying that the role of the superintendent in this is: some places you set something up, and step back because you know the people have to work it out. And then sometimes when you see it get stuck, then you go in and say, "Hey, this is the law."

RC: And it does not need to be a superintendent. It could be the deputy superintendent for instruction or the science coordinator. But I'm sort of unique. Instruction has always been my bag. I do all the other management and budget stuff because I like doing it. But I never forget that I was a teacher.

AS: So if I were to extrapolate from our experience should one approach the person that is in charge of instruction rather than the superintendent?

RC: Yes. But you need to get the superintendent to buy in.

AS: What about the role of the school board?

RC: You have to be careful about this. People can use projects like this for their own political agendas. School board members should be invited, should be aware of what is happening. But you don't want a board member that comes and says to a superintendent, "I want you to do this," and then the board member goes away, and then the superintendent changes, and the program goes away. What I've attempted is to do it so that you can change the board and the superintendent, and you can change some other people, it's never going to go away. It has become an intregal part of the program.

AS: What did you do as a superintendent to know that this was a program that you wanted to support?

RC: There is an ingredient here that I think has been good, that the age span of people that have come in contact with the kids has been very varied. And I think that's been very healthy. From young people like your medical students to young scientists to people my age and older.

AS: I guess what I was getting at is, I had the sense that you were going around talking to teachers and visiting schools to see what was going on. Otherwise you could just read some brochure that we print and accept that this is what we're doing.

RC: But that's just me. When I see something, I ask teachers, "How is it going? Is it helpful, are you using it?" Or, "Did that come from this summer's workshop?" One of the things that you have done is that you have brought adults together to learn. You have created a collegial community among our staff. That's very important. You have opened the doors of the classroom. We hear about isolation, and you have really helped that. I see teachers after the professional development helping each other. It was interesting when I visited the summer program that they insisted that I do it. I was hesitant, but they made me look through the microscope, "Well don't you see it? Look right there. Now, look over there, then you can see what is happening." I mean, the teachers were treating me like I want them to treat students.

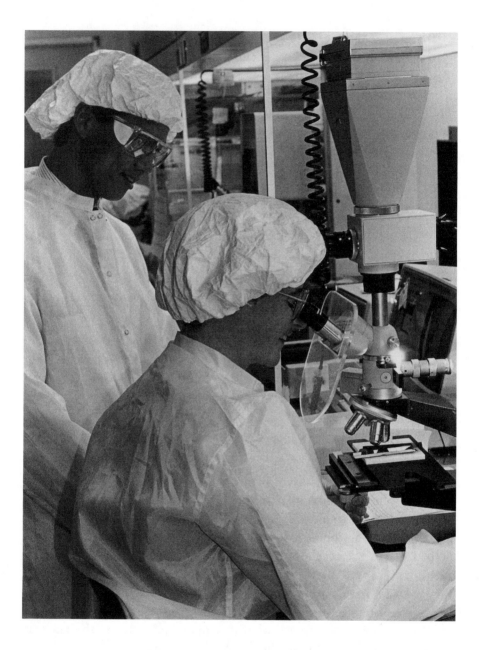

A teacher and industry mentor working in Raytheon's semiconductor fabrication laboratory as part of the Industry Initiatives for Science and Math Education program (photo courtesy of IISME).

Teacher Centered Activities

One-on-One Partnerships

BY DAVID STATES, SUSAN BRADY AND ART SUSSMAN

Perhaps the most direct way to bridge the abyss between precollege science educators and practicing scientists is simply to introduce these strangers to each other and provide a framework for their interactions. Our experience is that these personalized partnerships have distinct advantages and disadvantages, and that some fairly simple guidelines can help maximize their effectiveness. This article describes two programs that establish and nurture these one-on-one partnerships.

THE SEP MODEL

The Science and Health Education Partnership (SEP) of the University of California, San Francisco (UCSF) officially began with an open meeting attended by scientists and San Francisco teachers. After brief remarks by school district and university administrators, participants formed focus groups to explore the possibilities for collaboration. The formation of one-on-one partnerships between university scientists and school teachers emerged as a high priority to enlist scientists in the improvement of precollege science education.

The SEP staff has played a pivotal role in creating and nurturing these personalized partnerships. Our intent has been to create ongoing, independent, personal and professional matches between individual scientists or clinicians and teachers. We provide suggestions for partners to initiate the partnerships, as well as curriculum materials and other resources to facilitate their involvement, but the partners are encouraged to be creative and self-motivated. They choose the frequency and type of subsequent interaction. Through this activity, university volunteers share technical information, donate equipment or supplies, provide access to university resources (other scientists, the library, a well-equipped darkroom), give lectures, host laboratory tours, assist with hands-on activities in the classroom, and offer moral support to the teachers. Some partners interact on a weekly basis, while others meet only two or three times a year. A critical factor is that teachers must feel they have developed a genuine, personalized link to the university.

We recruit and place partners throughout the year, but this activity is most intense in the fall. Teachers and university volunteers are invited to an annual "Kick-off" event in early October. The Kick-off is advertised through school and university publications, the SEP newsletter, bulletin boards and word of mouth. The Kick-off highlights all program aspects, but we have often emphasized the partnership option by presenting exemplary partners, providing concrete suggestions that have been proven to work, demonstrating materials that partners might use in their collaborative efforts, and distributing sign-up sheets. It is important to set a relaxed, informal atmosphere for the Kick-off to enable people to break out of their usual cliques and meet new people. Simple ice-breaking activities help. At one Kick-off, we did a hands-on activity in which everyone had to shake hands with at least three new people during a specified five-minute time period. We provide refreshments to encourage people to relax and mingle.

The Kick-off provides a forum where potential partners have an opportunity to meet each other face to face. These direct encounters often prove critical to the formation of long-lasting partnerships. Partners who select one another tend to feel more comfortable with each other and are likely to overcome the inevitable awkwardness of initial phone contacts, and then connect in meaningful ways. The best partnerships seem to grow between people who find they share other personal outlooks. Many of our best partnerships eventually lead to close friendships.

For those who want to have a partner but have not made their own arrangements, the SEP program coordinator acts as a computer "dating" service to make the initial connections. In addition to the usual personal information (home and work addresses and phone numbers), the coordinator solicits a pool of critical details from the partner. These include subject areas of interest or need; grade and skill levels of the teacher's students; potential special language problems or cultural needs of the students; best contact times (e.g., teacher prep periods or the scientist's work schedule); the type of participation (classroom involvement, curriculum development, lectures, etc.); and areas of special expertise, teaching experience or access to special resources. Partners are asked whether they prefer male or female partners, or partners from specific ethnic groups (role modeling, for example, is often a critical factor). Potential partners need to provide an honest evaluation of how much time they are willing to commit. The more information provided, the more effectively the coordinator can arrange initial matchings.

Our approach to partnerships has evolved over the years. Originally, we emphasized setting up a large number of one-on-one partnerships by making educated guesses in matching people and then giving the scientist and teacher each other's phone numbers. With large numbers, we could provide only minimal guidance and follow-up. With time, we have chosen to emphasize quality over quantity. We are not rigid in terms of the structure but adapt continually to fill the needs expressed by teachers with the particular skills that volunteers make available. We recommend following the progress of partnerships more closely and nurturing them by providing materials, ideas and troubleshooting. To minimize disappointment, we bluntly warn everyone that personalized partnerships require a special commitment and we point out that there are many other program options for teachers and volunteers. We have also expanded the partnership concept to include teams in which several university volunteers might work with one teacher, or a volunteer may participate with two or more teachers on a special project.

CASE STUDIES

1. Teacher Bill Milestone met both of his partners — graduate student David Bowen and medical student Alfred Kuo — when he was brought by a teacher colleague to a SEP Kick-off. Bill was looking for scientists

to provide classroom assistance and role modeling for his students, many of whom speak English as a second language. Bill was not looking for technical assistance in a specific subject, but rather enthusiasm and willingness to interact with general biology students. He discussed the possibility of weekly visits with David and Alfred during the Kick-off and requested them as partners on our sign-up sheet. A SEP coordinator later confirmed that David and Alfred were equally interested in working with Bill. They were formally matched as partners, with Alfred, who speaks Cantonese, working on a weekly basis with the ESL students (predominantly Chinese-speaking) and David meeting with mainstream 10th-grade biology classes. Both university volunteers provided direct assistance with hands-on activities in the classroom. David and Alfred also used contacts in the SEP office and other labs to procure materials to facilitate the hands-on activities.

The liaison with Bill represented David's second attempt at a partnership. His first partnership with another teacher "never really got off the ground. The teacher didn't return my phone calls and I felt like I was always pushing my ideas too hard. I was looking for someone more enthusiastic, and Bill has been great. He's very organized and is really explicit about what he needs."

2. In 1987, Professor Roger Cooke was matched on paper as a partner to teacher Aljona Andrejeff. SEP staff initially viewed this partnership as something of a failure because of the infrequency of interactions (approximately three times a year). Despite this private diagnosis, Aljona and Roger quietly established their own criteria for success as they maintained their "infrequent," but steady, relationship over the last five years. Once a year, Roger organizes laboratory tours for Aljona's students. He visits Aljona's classes annually to present talks on muscles, or arranges for a friend to talk about Third World parasitic diseases with Aljona's class of predominantly immigrant students. Roger also independently solicits and donates occasional surplus materials. It was not until their third year that SEP staff recognized the success of this partnership. In short, Aljona and Roger had bypassed our office, and met each others' expectations. Most importantly, Aljona knows that when she needs something, be it technical advice or miscellaneous materials, she can contact Roger.

3. SEP staff assigned Professor Dick Shafer as a partner to Mission High School teachers Len Poli and Russ Janigian. After a slow start, the prodding of an SEP coordinator motivated the two teachers and UCSF professor to begin collaborating to increase student interest in the high school biology club. Although Dr. Shafer's time was somewhat restricted, the three managed to meet and develop innovative, hands-on molecular biology experiments. Students from the biology club even came on Saturdays to participate in the activities — a clear sign of genuine student interest.

The threesome also took advantage of Dick's faculty position to successfully apply for a prestigious University of California Presidential Grant for School Improvement (this funding is separate from SEP program money) in order to expand and improve the units. They received $15,000 to inject life into Mission High School's science curriculum. Len and Russ were subsequently chosen by the school district to play a major role in organizing and designing new hands-on biotechnology units for the district at large. Not surprisingly, they have harnessed other SEP resources to carry out this latest endeavor, including soliciting the participation of four UCSF volunteers as technical advisors and acquiring a large amount of donated equipment for the project.

4. Postdoctoral fellow Steve Doxsey met middle school teacher Gerry Pelletier three years ago at a SEP Kick-off. Their partnership struggled initially because telephone calls did not connect and work schedules did not seem to mesh well. Finally, they found a formula that worked by taking advantage of another SEP activity as a vehicle for their collaboration. Gerry enlisted Steve to advise his 7th- and 8th-grade students with their entries in the SEP Science and Health Lesson Plan Contest, an annual opportunity for students to compete for prizes by teaching science or health lessons to their peers or younger students (see the contest article on p.157 for details). With a defined objective, Steve began to help Gerry's students develop their ideas about hands-on lessons and critiqued preliminary presentations. Steve's easy manner with the kids, love of science, and confidence made him a great role model. The students won first prize in the middle school division that year and were subsequently highlighted in an article in Science magazine.

These represent only a few types of successful partnerships. There is no set formula. In fact, we view the diversity of partnerships as being healthy and desirable; we encourage people to be creative and find their niche, however large or small. Ideas abound and new ones are generated every day — the trick is to link creative people who have similar desires and expectations, and then find ways to encourage these many diverse partnerships.

THE STEP MODEL

Funded by the National Science Foundation, the Scientists and Teachers in Educational Partnerships (STEP) program builds partnerships among individual scientists, teachers and students by means of classroom activities that introduce societal issues related to recent advances in science. In its first three years, nearly 250 scientists from industry, universities, medical groups and government agencies and teachers from Northern California high schools formed STEP partnerships. The Lawrence Hall of Science (LHS) administers the project.

The STEP program grew out of discussions with educators and scientists concerned with the quality of science education in California. Teachers expressed a need for help in bringing into the classroom the content of current scientific advances, and for assistance in providing more laboratory experiences for their students. Scientists expressed a desire and willingness to be involved in education, but many were unsure of how to make use of their skills and knowledge in a classroom setting. Both groups were frustrated by a lack of strategy for the formation and maintenance of partnerships that make the best use of their individual strengths and abilities.

Coordinators from the Lawrence Hall of Science met with a variety of university, industry and educational leaders to plan a strategy. Unlike the UCSF partnership model, STEP planned to connect participants working in many different scientific and educational institutions. STEP coordinators, therefore, decided to provide a more structured framework for the teacher-scientist partnerships. The general program strategy is to train scientists and teachers together in specific hands-on science activities related to societal issues as a means for establishing and maintaining relationships with each other. Partners then co-teach in the classroom, developing additional activities that introduce students to specific areas of research.

THE STEP PROGRAM INCLUDES:

• A one-day Educational Inservice early in the school year in which scientists and teachers learn an introductory, hands-on LHS activity and generate ideas for modifications and extensions. Topic areas include DNA Science, Drugs and the Brain, Global Warming, Plastics, and AIDS Epidemiology. Partnerships are formed and strengthened at this event. Activities and strategies for introducing societal issues in the science classroom are also presented.

• Classroom visits during the school year. Scientists and teachers implement the core lessons in which they have been trained. They also work with LHS staff to develop and co-teach extension activities to communicate the scientist's area of specialization. LHS staff monitor partnerships throughout the year and provide materials and teaching assistance.

• The High School Science and Societal Issues Symposium, an annual LHS event designed to motivate high school students to grapple with ethical and societal issues raised by recent advances in science and technology. Students make presentations that incorporate the scientific and ethical content issues. Scientists act as information resources for students and they also judge the presentations.

• A Spring Follow-up Meeting, designed to enable participants to assess the progress of the project and to generate new ideas for activities and improvements.

Teachers and scientists are selected for partnerships based on several criteria. They must teach in close geographic proximity to available partners to minimize commute time and to build local commitments between schools and science-rich institutions. Teachers who are bringing a student team to the LHS Science and Societal Issues Symposium receive top priority for partnerships, and it is expected that scientists will play an important role in helping students prepare their presentations.

During the Fall Educational Inservice Day, the scientists and teachers participate in a specific activity. LHS instructors model activity-based science by using the same techniques for data collection, analysis and discussion of the results that scientists are expected to use with the students. The instructor comments on class management techniques and models how to lead discussions about the concepts that are being communicated through the activities. In addition, activities and strategies for addressing the controversial societal and ethical issues raised by advances in science are presented in order to increase the confidence of teachers and scientists in dealing with these issues in the classroom. Participants also work with LHS staff to learn procedures for equipment kit loans, scheduling of visits, and details about the Science and Societal Issues Symposium.

Scientists make at least a four-day commitment to classroom visits during the ensuing school year. The major task for the scientist and teacher is to build a relationship with each other and then involve the students. The scientists also serve as resources for the student teams who are preparing a presentation for the Symposium. One purpose of the STEP program is to help scientists and teachers implement hands-on LHS activities so they both gain confidence and will be able to lead that particular activity as well as other laboratory-based programs with students in subsequent years. Participants have access to LHS equipment and supplies for teaching classroom activities and extensions, and STEP staff are available throughout the year to assist in the development and modification of activities by scientists.

The model science education materials are chosen from several LHS instructional series developed for middle school and high school students. All LHS materials encourage hands-on, guided discovery of scientific concepts, promote the interaction of small groups through cooperative learning, and foster the development and use of critical thinking skills. Each activity has been tested with a wide range of students and teachers, and has been modified based on feedback from participants. STEP staff have extensive experience in the development and modification of educational materials and in conducting workshops that enable teachers to use activities effectively with students.

Scientists who learn the same LHS activity may develop completely different extension activities, depending on their particular areas of expertise and interest. For example, a scientist who does molecular genetic research on plants may select the DNA Science activity. This scientist might develop extension activities that demonstrate how the techniques used in DNA fingerprinting can be used to isolate and insert specific genes into plants. Another scientist may be involved in forensic work, and would develop extensions that relate to the use of DNA fingerprinting in legal cases. A genetic counselor might use DNA Science as a way to introduce the use of differences between restriction enzyme patterns as a way to diagnose genetic disease. It is this exciting process of developing extensions that brings the scientists' particular areas of specialization alive for teachers and students. Following are descriptions of two of the model LHS activities:

Global Warming and the Greenhouse Effect

Global Warming and the Greenhouse Effect is a teacher's guide in the GEMS series. The students explore this complicated and controversial environmental topic by building a "greenhouse model" of the atmosphere using light bulbs, bottles and thermometers. After discussing the limitations of their first greenhouse model, the students go on to play the "greenhouse game," a conceptual model that illustrates how carbon dioxide in the atmosphere contributes to warming Earth by absorbing infrared photons. In other sessions, students learn ways to detect the amount of carbon dioxide in a gas sample, then experiment to compare the amounts of CO_2 in car exhaust, human breath, air (from a hand pump), and the gas created by the reaction of vinegar with baking soda. Students also analyze scientific data and articles from varying viewpoints, including a graph showing the amounts of CO_2 contributed to the atmosphere by different world regions.

AIDS Epidemiology

The AIDS Epidemiology simulation is a hands-on activity that uses this disease as a model for teaching about the use of statistical data in epidemiological research. Teachers initially poll students on their attitudes about AIDS in preparation for a role-playing activity that mimics the spread of the disease. Students are assigned various characterisitcs by rolling dice. They then simulate sexual contact by selecting beans from a series of paper bags that represent different demographic groups.

Throughout the simulation, small groups of students use manipulative materials, generate and collect data, and record the results of their actions. The class comes back together to pool their results. The data collected by the larger group are used as a springboard for examining how the risk of spreading AIDS is determined by the behavior of each individual. The results for specific population groups mirrors the epidemiological data concerning the spread of the disease. This information is a catalyst for thoughtful and informed discussion on a controversial issue.

As is true for the UCSF program, STEP staff emphasize individuality and creativity in designing their particular partnership. The four days of classroom interaction may be the same activity done four times for different classes, or it may be four different activities. Scientists and teachers may lead LHS activities or they may design their own. Students may visit a scientist's lab or field site. Many scientists provide their teacher partners with supplies and/or equipment. Some teachers and students work during the summer at the companies or institutions that sponsor their scientist partner. As we share feedback among the participants and across the years, the partnerships continue to grow and develop.

HELPFUL SUGGESTIONS ON HOW TO FACILITATE PARTNERSHIPS

1. **Emphasize quality vs. quantity in partnerships.** Many teachers and scientists may volunteer for this type of activity because it sounds interesting, but they often do not really know what they are getting into. Make sure they know. Recruit participants carefully. There is no sense in placing people who really cannot make this type of commitment. They and their partners will only be disappointed in the end. A certain percentage of partners, especially at the beginning, will need support in terms of practical ideas, ways to overcome problems, or assistance in sorting out occasional misunderstandings. Twenty-five to 40 partnerships is probably the effective limit for one coordinator to monitor.

2. Exploit special events to bring potential partners together.
Personal chemistry is often crucial in forming partnerships. Potential partners can communicate directly far more information about what they need than they can through a broker. Initial face-to-face encounters are also often the key to encouraging partners to pursue each other. Leverage other program activities (Lesson Plan Contest, Societal Issues Symposium) to increase partnership interactions.

3. Gather as much information about participants as possible. If a coordinator must make a "paper" match, it is crucial to pair teachers and professionals with similar expectations. Maintain records in an easily searchable database such as Filemaker Pro.

4. Provide potential materials and/or ideas for partners to use.
Especially in the beginning, many partners are not sure how to work with each other. Encourage partners to exploit other program opportunities, e.g., speakers' bureaus, contests, tours, etc. STEP has experienced success with providing very structured activities for the partners. SEP has the benefits of diversity but its less structured approach may make it more difficult for some partnerships to get off the ground.

5. Insist that assigned partners meet as soon as possible.
Procrastination is a deadly sin. Take the volunteer to the school site for an initial meeting with her partner, if you have the time. Telephone tag is the most frequent cause of partnership death. Encourage both partners to call each other and be assertive. Teachers, as well as science and health professionals, are difficult to reach. This is often misinterpreted as disinterest. Warn partners about this problem and tell them to be persistent. Get home phone numbers since it is often best to make contact in the evenings or weekends.

6. Follow-up partner progress on a regular basis, once every two or three months, for example, to make sure things are going smoothly. When problems exist, frustrated partners rarely call a coordinator to sort them out. Instead, they usually procrastinate and the partnership withers. The sooner the coordinator knows about problems, the sooner they can be resolved.

Conclusions

Theory is often simpler than practice, and a good number of attempted partnerships fizzle out. Individualized partnerships are not for everyone. Experience has taught us that this type of independent activity demands a special type of self-motivated participant. Nevertheless, we believe that the potential gains can far outweigh the difficulties.

Teachers, scientists and staff have found that the one-on-one partnership is an effective way to introduce more current science into the classroom. In addition, such relationships encourage leadership by scientists to establish partnerships with local schools and to provide more laboratory experiences for students. Partnership activities also give students a golden opportunity to learn more about careers in science and technology.

For teachers, partnerships provide a range of rewards including assistance in the classroom, access to technical expertise and material resources, and a source of professional role models for their students. More importantly, the morale building that comes with special treatment of the teachers by respected university and industry scientists and health professional has an incalculable positive effect in countering "teacher burn-out." Successful partnerships are a palpable demonstration of our concerted efforts to work with teachers as professional colleagues in the science and health community.

Interestingly enough, scientist volunteers claim to receive as many rewards from their collaborations as their teacher partners. The personal relationships that are built during the exchange, both with teachers and students, cannot be underestimated. The satisfaction of escaping from the pressures of research for a few hours to teach and interact with young people is often cited as a prime reward for participation. Many scientists genuinely miss the opportunity to share their knowledge with young people. In the process, scientist partners, especially those who visit the schools, learn first-hand the difficult circumstances in which our teachers are expected to operate. These experiences dramatically reinforce the importance of providing assistance. Perhaps the greatest satisfaction comes from finally having an opportunity to actually do something that is both positive and enjoyable to improve science education.

ONE PARTNER'S STORY

BY MIKE MORGAN

Five years of graduate school gave me the research skills necessary to be a scientist and the teaching skills necessary to teach at the college level. This education, however, did not prepare me to teach in public schools. The Science Education Partnership (SEP) has allowed me to get around this deficiency and, thus, have a very productive and rewarding teaching experience in the public schools of San Francisco.

I received my undergraduate and graduate degrees from the Department of Psychology at the University of California in Los Angeles. Currently, I am doing postdoctoral research in the Department of Neurology at UCSF. I have held this position for two and a half years and have been a volunteer with SEP throughout that time. My teaching experience before becoming involved in SEP included working as a teaching assistant for four years at UCLA and as an instructor for a year at Santa Monica College.

As a SEP volunteer, I have worked with students in the second, sixth, seventh and 10th grades. Working with students at different education levels made me aware that many students lose interest in education as they move from the second to the 10th grade. Every student in the second grade wants to answer questions and please the teacher. In the sixth and seventh grades most of the students still enjoy learning, although a few students have begun to drop out mentally. By the time students reach the 10th grade, there are two distinct groups: those who are interested in education and those who are not. This second group made my volunteer work in some high school classes seem useless. At the very least, working with students who did not want to be in class was a challenge beyond the scope of my semi-weekly visits.

During my first year as a SEP volunteer, I worked with a physical education instructor who wanted help with three 10th-grade basic biology classes. The "personality" of the classes ranged from apathetic to rude and disrespectful. I expected that the presence of a scientist in the classroom would be of some interest to the students or, at the least, that I would be shown some respect. I was wrong. Some of the students never lifted their faces from their desk, others never stopped talking, and a few walked around and out of the classroom whenever they felt like it. Discipline was not something that I had thought about or discussed with the teacher before entering the classroom. I would end up hoarse after three classes of talking to students who did not want to listen. Needless to say, this was a very frustrating experience. The difficulty in teaching these students is best conveyed by a statement made to the teacher by one of the better behaved students: "I hate school. I don't need to learn this. I'm not going to college." It was clear that most of these students did not want to be in school.

I witnessed a much different environment in an advanced 10th-grade biology class. I spent only one day as an observer in that class, but it was enough to make me aware of the extreme difference between the basic biology class and the advanced. These students were interested in learning and could carry out activities with little direct supervision. I could see that here the educational process was working. It was also obvious that the students in the basic biology class, for whatever reason, had not taken advantage of the educational system. I didn't feel that I had much to offer either group. The advanced students were already interested in science and the basic students did not seem to care whether I was there. At the end of the year I decided to try my luck with a younger and more impressionable group of students — sixth- and seventh-graders.

At the middle school in which I currently volunteer, even sixth-grade students are divided into advanced and basic science classes. Although there is an obvious difference between these classes, the difference does not preclude presenting the same lesson to both groups. My partnership is with a young and enthusiastic biology teacher. She teaches a combination of sixth- and seventh-grade classes, some of which are advanced, others basic, and some for students for whom English is a second language (ESL). During each visit I try to attend all five of the classes that comprise a teacher's daily classload. This is exhausting, but I feel that it is the best use of my time. I try to get to the school once every two weeks, although sometimes I attend on consecutive weeks, other times once a month.

I always do my homework before attending a day of classes. This involves talking with the teacher to find out what each class is studying and determining if and how my presentation will relate to these lessons. My preparation also may include reviewing the topic of study, writing a lecture, or practicing an experiment or demonstration. Preparation can be difficult when so many different groups of students are encountered in one day. I try to adjust my presentation to fit the needs of each class. For example, I use the chalkboard more often in the ESL classes so as to help the students with vocabulary.

There are many different ways that I have assisted these classes. I have presented lectures, led discussions, done demonstrations, helped students do experiments, assisted with ongoing assignments, and spent the day answering students' questions. My presentation is determined by what the students are studying, the availability of relevant materials for a demonstration or experiment (the SEP office is of great help here), and the amount of time that I have to prepare. If my work schedule is very busy I still make an effort to attend class, but may only assist with the teacher's ongoing lesson so as to reduce my preparation time. When I have time to prepare a demonstration, I try to make the presentation relate to the unit being studied in class, although this is not necessarily so. Since my scientific specialty is the relationship between the brain and behavior, I tend to gear my presentations to this topic. One of the most enthusiastically received presentations was when I brought a human brain to class. I began with a brief description of neuroscience and the function of the brain, and then divided the students into smaller groups so they could approach the brain and ask me questions. All of the students had questions and nearly every student wanted to touch and hold the brain. The excitement resulted in a surprisingly good memory of brain parts and function.

I am not afraid to present topics that are outside my area of specialty, however. A little preparation allows most topics to be mastered well enough to present to sixth-grade students. Moreover, I do not mind the students teaching me about a topic that they are studying, although this tends to surprise them. They often interpret a question from me as if

I'm testing them rather than the possibility that I may not know the answer. I think it is important for visiting scientists to admit that they do not know everything. It underscores the scientific process of asking questions, provides an opportunity for the class to learn something together, and makes the scientist seem "more normal." Learning is fun when accompanied by a sense of discovery and not when all of the answers are provided by an "expert." A classroom of 30 students can come up with some very reasonable hypotheses to some very difficult questions.

I had many questions and concerns before entering the classroom as a SEP volunteer for the first time. How do I interact with sixth-graders? Will the students cooperate? Will they understand what I'm saying and find the lesson interesting? What does the teacher expect of me? These questions and others were answered by experiences in the classroom. Each visit became easier and more enjoyable as I became more familiar

---

I think it is important for visiting scientists to admit that they do not know everything. It underscores the scientific process of asking questions, provides an opportunity for the class to learn something together, and makes the scientist seem "more normal."

---

and comfortable with each class. It is a good feeling to have students enter the class with a "Hi Mike. What are we doing today?" There is no substitute for the time it takes a scientist and teacher to get to know each other and define what they expect from each other. I entered the class without knowing what sixth-graders could do and how the partnership was supposed to work. I found out that the partnership works however the teacher and scientist develop it.

Discipline is a potential problem that I had not thought about before entering the class. Each teacher has techniques that they use to discipline their students, and, thus, discipline is something that I prefer to leave to the teacher. Nonetheless, because I have worked with the same teacher for a year and a half, I am now much less hesitant to discipline students in a manner consistent with her techniques. Both the teacher and I enjoy interacting with the students, so we accept some unnecessary noise as the cost for this interaction.

My experiences as a SEP volunteer have included visits to a few second-grade classes. Like the sixth-grade students, second-graders are a joy to work with. They are excited just to have someone in their classroom. They steal looks at you as you wait in the back of the class to be introduced. Every student wants to answer every question. In fact, I learned rather quickly that calling on a child with a raised hand does not mean that the student has anything to say. It is difficult to say whether students this age get much out of these visits — three students asked me the same question in less than 10 minutes. I think just being in the class as a representative scientist was more important to the students than anything that I said.

I enjoy working with people who want to learn. This is true whether these people are in second, sixth, seventh, 10th grade, or college. Last summer, I spent an enjoyable and intense month teaching science to second-grade teachers. My preference, however, is working with sixth-grade students. They are old enough to think on their own and process new information, yet they are not too old to have become apathetic about education.

My motivations for devoting time to volunteer in the schools are many. These include: Combating negative stereotypes about scientists created by the movie industry and animal rights activists; impressing upon students that learning can be fun and that it is important; and, in a small way, enhancing science literacy. In addition, I hope that the time I spend in the classroom allows the teacher to help more students and provides the encouragement that some children may need to succeed in school.

INDUSTRY INITIATIVES FOR SCIENCE AND MATH EDUCATION

BY BRIAN KEARNEY, MARIE EARL, KARIN ROSMAN,
KATHRYN SLOANE, ZINA SEGRE, WILLIAM ALLSTETTER

Industry Initiatives for Science and Math Education (IISME) was founded in 1985 by a consortium of San Francisco Bay Area companies and government laboratories in partnership with the Lawrence Hall of Science (LHS) at the University of California at Berkeley. IISME was founded to address the critical need for a strong, highly skilled workforce in mathematics, science and other technological fields. Paid summer internships for teachers in industry form the core of the IISME model. Industry partners provide the summer work positions and unique access to resources, experience and insight into applying math and science concepts. LHS provides the expertise and credibility that are instrumental in translating the summer experience into improved classroom instruction.

Gareth Wong has spent his entire life in school, first as a student, then 20 years as a high school chemistry teacher. But in the course of three summers he helped synthesize a biological marker for Syntex, tested compounds that lubricate IBM disc drives, and helped Lockheed uncover a glitch in the way it manufactured protective tiles for the space shuttle. More than just jobs, these experiences exposed him to a new world and they changed the way he teaches.

"These jobs had a definite impact on my teaching," says Wong. "I have seen what industry does on a day-to-day basis and can relate that experience to my students. I've gained more credibility from the fact that I've been there and seen what industry wants, what the needs are. The experience also increased my self-confidence. I found out I could do quite well because a lot of things I do as a teacher carry over to the industry environment."

This is exactly what the founders of Industry Initiatives for Science and Math Education (IISME) had in mind. IISME is a partnership of nearly 60 San Francisco Bay Area companies and the Lawrence Hall of Science. It seeks to improve math and science education primarily through paid summer fellowships for teachers. Teachers learn the needs and expectations of industry, as well as its opportunities. They take that knowledge, and often a renewed enthusiasm for teaching, back to the classroom.

The job opportunities are as varied as the teachers who fill them. Mark Meeks helped Dow increase solvent recovery from waste water. Gary Hensley directed a video. Jean Duda wrote a Hypercard program to display fiberoptic and telephone products for Raynet.

> **The goal of the summer work experience is to empower teachers by providing them with experiences and resources that will enrich the classroom experience of their students. The summer job is not an end in itself. When teachers take back to their classrooms new ideas and insights about their subjects, enthusiasm for trying new teaching strategies, up-to-date information for use in counseling students, and strong relationships with industry sponsors — then IISME has achieved its goals.**

In the first eight years of the program, sponsor companies, hospitals and government laboratories have offered nearly 600 summer fellowships to teachers in the seven-county Bay Area. The more than 350 teachers who received these fellowships represent approximately 15% of the Bay Area high school science and mathematics teaching force. Through these teachers, IISME-related ideas have reached over

300,000 students. Industry sponsors have contributed close to $5 million to improving mathematics and science education in the Bay Area in the first eight years of the program. In addition, the sponsors have contributed over 8,000 volunteer hours as mentors to teachers, coordinators of the program within companies, guest lecturers in classrooms, hosts for company tours, and counselors to IISME and to schools.

Host companies can hire teachers through IISME (now established as a nonprofit tax-exempt public benefit corporation) or enroll them directly on company payrolls. The cost to the companies is the teacher's salary ($700 per week) and IISME program costs ($275 or $375 per week depending on whether IISME administers the payroll). Since the work period lasts eight weeks, the cost per teacher is approximately $8,200. Corporate, private and government foundations also provide support for general operating expenses, academic year programs, and replication of the IISME model.

IISME staff administer the program, including industry recruitment, teacher placement and payment, educational follow-up, program replication, public relations and special projects. A Senior Advisory Council — consisting of six senior-level executives of sponsor companies, the chairman of the Lawrence Hall of Science, and an educational leader from a local school district — provide general guidance and assist with resource development. A board of directors — consisting of the director of the Lawrence Hall of Science and local educational and business leaders — determines IISME policies and procedures.

Within participating companies, an on-site company coordinator serves as the primary liaison between the IISME staff and company personnel. The coordinator's responsibilities include identifying industry mentors (the people who work directly with the teacher during the summer), overseeing the development of summer jobs, distributing IISME materials and information, coordinating fiscal and other administrative tasks, and assisting with the teacher interview and selection process. Most coordinators also arrange for teachers to attend tours, meetings and training classes, and provide access to other company resources. The industry mentors make a substantial time commitment to work closely with and provide guidance to the teacher fellows during the summer. They are often active in academic year follow-up as well, visiting schools or inviting students to their industry sites. One of the most difficult tasks is finding the bench workers willing to make the necessary commitment to ensure a mutually successful experience.

In 1991, IISME for the first time met its goal of placing 100 teachers in industry summer jobs. The roster of companies hiring teachers has varied each summer, usually reflecting economic conditions. A core group of companies has provided a steady pool of job opportunities. Teacher placements within companies range from a single teacher to as many as 15 at the Bay Area's larger employers.

Until recently, IISME had accepted applications only from high school mathematics, science and computer science teachers. For the past two summers, IISME has piloted offering the program to teachers in grades 6-8 as well. IISME only hires teachers who have completed at least two years of full-time teaching. We believe teachers who have less classroom experience typically have more difficulty achieving the required industry-to-classroom transfer.

Participating high school teachers generally hold undergraduate or higher degrees in a science- or mathematics-related field, making them desirable employees for sponsor companies. Increasingly, companies are indicating a willingness to hire middle and junior high school teachers (who generally do not have the same level of academic preparation). These companies recognize that students often decide before they reach high school whether to pursue higher levels of mathematics and science course work.

Teacher recruitment begins in September for the following summer. IISME mails comprehensive information packets to every vice principal, mathematics department chair, and science department chair in the Bay Area. In addition, using National Science Teachers Association mailing labels, application-request flyers are mailed directly to teachers at their school addresses. In all, IISME mails approximately 4,000 information packets for teacher recruitment.

Then begins the hard job of prioritizing applicants and matching teachers to jobs. The IISME Lawrence Hall of Science office reviews each application and assigns a rating code based on a variety of education-related criteria. These include previous professional development experiences, communication skills, leadership potential, ethnicity and student population (especially with respect to at-risk categories), and participation in extracurricular activities and enrichment programs.

Job matching proceeds in the order in which job descriptions are received at the IISME office. The teachers' educational background, computer skills, and willingness to commute to a job site are the primary factors determining whether teachers will be placed. Teachers are notified of referrals at the time IISME mails their applications to industry sponsors. Mentors are urged to screen candidates carefully and to conduct personal interviews, evaluating the teachers as they would any new employee. In the early years of the IISME program, candidates were referred to only one company at a time. For the last several years, teacher applications have been submitted to several companies at the same time. This "free market" referral system spurs mentors to interview and make hiring decisions as quickly as possible. However, the job match process routinely continues until the second week of June.

IISME coordinators make weekly telephone calls to company contacts to track interviews and hiring decisions. By late April, the pressure to bring closure to hiring decisions increases greatly. Some applicants withdraw from the program if a promising job is not on the horizon or if other possibilities materialize. Job matching often continues into the second week of June, just before the fellowship period begins. Each year approximately 20 jobs have been withdrawn by industry sponsors, often late in the job match period. The reasons for job withdrawal vary greatly, ranging from changes in mentor assignments and budgets to requests for lab or computer skills only rarely available among IISME applicants.

IISME allows teachers to work at the same company for no more than two consecutive years. This policy is designed to provide more diverse work experiences for teachers, to prevent teachers from developing "ownership" of particular jobs, and to allow greater numbers of teachers to participate in the program. The IISME Board of Directors has declined to place any other limits on repeat participation since the teachers, IISME and the companies all benefit from the insights and experience contributed by returning teachers.

Of course, the ultimate goal of all IISME efforts is to enhance classroom mathematics and science education. To that end, IISME's Education Office at the Lawrence Hall of Science sponsors a number of activities and provides academic support to Fellows during both the summer and the school year. During the summer fellowship period, there are three meetings held at the beginning, middle, and end of the eight-week session. These meetings help orient new participants to the goals, mechanisms and expectations of the program; plan the transfer of concepts and resources from industry to the classroom; and celebrate the successful completion of the summer period.

One returning IISME fellow has the job of peer advisor to the fellows, visiting them at their industry sites. The peer advisor facilitates communication between fellows at different sites, consults with fellows on Action Plans (see below), assists first-time fellows who need additional support, and acts as a general trouble-shooter. This peer advisor position was originally funded by a grant from the National Science Foundation, and yielded such positive benefits that the position is now funded by an industry sponsor each year.

Fellows are encouraged to meet regularly over the summer at their company sites. Sites with only one or two fellows may join a group at a neighboring site. Some groups meet informally, while others choose to meet for scheduled activities. These meetings reduce feelings of isolation among fellows (reported in the earlier years of the program) and provide fellows with an important informal, collaborative setting in which to explore the educational implications of their industry experience.

Each teacher fellow is responsible for producing a plan for the classroom transfer of some aspect of their IISME experience. In the early years of the program, teacher fellows were required to submit a curriculum project at the end of the summer describing lessons they had developed for their classroom based on their industry experiences. However, this approach did not work well. Many teachers worked on job assignments that did not directly apply to the content or level of the courses they taught. Developing new curriculum materials required considerable time (difficult to find during full-time summer employment) as well as time to pilot, test and revise. In addition, only a minority of the fellows were experienced in or felt comfortable with writing curricula, and the quality of the resulting projects was inconsistent. Many teachers also felt that the focus on content overlooked many of the salient and important implications that the industry experience had for instruction. Finally, unlike traditional teacher workshops, the approximately 100 IISME fellows work at many different sites and have a wide range of experiences, so there need to be a variety of options for translating these experiences for the classroom.

Teachers emphasized that the industry experience motivated them to place increased importance on instructional strategies that encourage teamwork, communication, problem-solving and good work habits. These were the vital skills that they noticed in their job environment and these were the skills that industry identified as being particularly important in preparing the future workforce. Teachers recognized that these skills were not traditionally emphasized in their classroom instruction.

As a result, IISME staff, with input from a team of experienced fellows, revised and modified the classroom transfer requirement. Instead of the curriculum development model, teacher fellows are now allotted 10% of their time over the course of the summer fellowship to develop an Action Plan. This Action Plan aims to transfer an important aspect of the industry experience to the classroom. It can focus on teamwork, problem-solving, curriculum lesson plans, career awareness, fostering increased school-industry interactions, or taking a leadership role in educational reform and improvement.

Industry mentors are encouraged to participate in developing and implementing the Action Plan. One industry mentor even developed a corporate version of an Action Plan. She realized that companies often provide in-service educational opportunities for their workers and that available openings in these training programs could simply be given to teachers. Her work has resulted in a catalog of educational opportunities for teachers and school administrators offered by Bay Area employers. Offerings include computer skills, management techniques, career enhancement skills and interpersonal relations.

All teacher participants also become permanent members of the IISME Academy, which hosts four meetings during each school year. Academies are designed to update fellows' content knowledge and to bring the IISME community together to discuss obstacles and strategies for implementing Action Plans.

When IISME was first established, many corporate sponsors viewed the summer fellowship as a benefit the companies provided for teachers.

Few saw the potential benefits to the company. Over the years, however, corporate sponsors have become quite enthusiastic about the two-way exchange of expertise and ideas resulting from the partnership.

> *"I absolutely think this is one of the best programs," says Mary Clifford, environmental services administrator at Syntex and an IISME mentor. "It is so well organized. It meets our business needs, it meets the teachers' needs, and the bottom line is that the students reap the benefits."*

The most immediate benefits to industry are the contributions and accomplishments of the teachers during their summer work assignments. Mentor evaluations of their teacher fellows' job performances are overwhelmingly positive each year. Fellows are perceived as being highly organized, self-motivated and capable in establishing objectives and meeting deadlines. Teachers approach their summer assignments with initiative, excellent communication and organizational skills, and participate as valued team members.

Mentors also appreciate the new perspectives and abundant enthusiasm that teachers bring to the workplace. Many employers find that their participation in the IISME program provides a morale boost for employees, who are impressed by their company's commitment to the community.

The primary concern and long-term benefit is that teachers return to classrooms with the skills and insights needed to prepare their students to be members of a scientifically literate workforce. The mentors are also encouraged to enhance the classroom transfer. Over 85% of mentors visit their teachers' schools and classrooms, host tours for students, or become directly involved in planning and implementing classroom activities. In addition to providing role models and unique experiences and resources for students, these connections enable industry personnel to gain a better understanding of the demands and constraints that teachers face.

What about the teachers? Year after year, over 90% of IISME teacher fellows rate IISME as one of the best professional development experiences available to them. Perhaps the most notable result of the summer in industry is the impact on teachers' view of themselves as professionals. They overwhelmingly rank "increased self-confidence" as one of the most important outcomes. Successfully completing a challenging job in their academic field, being treated as a professional by their industry colleagues, discovering they have the skills and knowledge to perform in a completely different environment — all contribute to an increased sense of self-confidence and self-worth. Contrary to early concerns that teachers would flock to industry jobs after their summer fellowships, most teachers feel that their summer experience revitalizes them and renews their commitment to teaching as a profession. In a 1991 follow-up survey of all former IISME fellows, 40% reported that the IISME experience encouraged them to remain in the teaching profession. Their net attrition rate was lower than the estimated California rate.

This enhanced professional self-confidence stimulated 70% of the teacher fellows to pursue new professional development activities and tackle new challenges. These included:

- Improving classroom instruction;
- Assuming new leadership roles in the school or district;
- Developing or redesigning curriculum;
- Assuming new professional responsibilities (e.g., department chair);
- Developing new computer curriculum or enrichment programs;
- Participating more actively in school restructuring or other education-reform efforts.

A common outcome reported by teachers is a better understanding of the skills needed to succeed in the industry environment. They gain specific and substantial information about careers, the skills needed to successfully work in mathematics and science careers, and the educational qualifications needed for these careers. This type of information is not generally available in a user-friendly form for teachers, as noted by the 35% of all teacher fellows who listed "knowledge of industry" as the primary benefit of IISME participation. The knowledge of industry and careers gained during the summer represents the set of information that teachers are perhaps most eager to share with their students. And,

WHO SAYS IT'S A MAN'S WORLD

It's a man's world. At least that's what many people think about careers in math and science. But when IISME Fellow Barbara Rodrigues spent a summer working at Syntex Corporation, she kept meeting professional women with math and science backgrounds. They worked at a wide variety of jobs, which they enjoyed and found fulfilling. As a math teacher at Mercy High School, an all-girl's school in Burlingame, she thought these women might have something to say to her students.

When she returned to school in the fall, she arranged for 31 girls from her algebra/trigonometry class to meet 13 women from Syntex. When they assembled at the company's Palo Alto office, the women introduced themselves and talked about their various jobs. The group then began a roundtable discussion about general skills needed in the workplace. Many of the women emphasized public speaking and negotiation. They also encouraged the girls to keep informed about current events and acquire a broad education that included courses beyond one narrow expertise.

After the discussion each woman took two or three students back to her workstation and explained the details of her daily work. Biostatistician Jill Fujisaki emphasized the role of math in her job. "I think they were surprised at how important it was. They didn't have any idea how math was used on the job. They saw how I use numbers every day, how I examine data, and how I analyze something on the computer."

Student Rosette Reyes looked at blueprints instead of a computer — and was thrilled. Reyes hopes to be an architect and was encouraged by her session with architect Christine Day. She also absorbed advice Day offered about a key to success. "I really enjoyed it," said Reyes. "That's really what I want to do but I can't just say that I want the job and expect to get it. You need to be aggressive. If you really want it you have to stick with it, really work at it and don't give up."

Other students learned about careers in toxicology, marketing, environmental engineering, regulatory compliance and human resources.

Back at school the girls explained what they had learned to other students with a variety of presentations that ranged from videos, to posters, to mock newspapers. Barbara Rodrigues believes the girls take their studies more seriously after the Syntex visit, especially if they can see a connection to a potential career. "The career awareness is so valuable in motivating them to continue their studies in math and science," Rodrigues concluded.

given the dearth of career counselors and career information available to students, this area is of considerable interest to industry sponsors.

This knowledge of careers and the education needed to pursue those careers provides one of the most direct transfers from the summer experience to the classroom. Teachers focus on the technical skills (such as "the absolutely universal requirement of computer literacy") students will need for future jobs. Or teachers draw upon their experience to offer advice to students about the importance of "lifelong learning." One teacher, for example, felt it was important to make students aware of the importance of risk-taking.

As mentioned previously, IISME fellows' heightened awareness of the importance of teamwork and of good communication skills, work habits and problem-solving abilities in the workplace encourages them to emphasize these skills more in their classrooms. These priorities are visible in more student oral and written presentations, increased use of cooperative learning situations, and reinforced emphasis on promptness, self-discipline and completing assignments on time.

While curriculum development is no longer the primary academic-year goal, many teachers do report using their industry experience to update course content. They often include examples from industry to illustrate concepts. As many teachers have said, "Now I finally have an answer to the age-old question, 'Why do I have to learn this?'" Using current examples of how concepts and skills are really used makes the subject more interesting and relevant to students. They have seen how science and math are used in the working world and can give examples of mathematical integration, oxidation/reduction reactions and momentum.

Many teachers also draw upon their summer work to revise the content of lectures and laboratory activities. For example, early in the IISME program, chemistry teachers recognized the need to introduce the topic of polymers into their high school courses. When possible, teachers also incorporate more technology into their curriculum. Some teachers describe the laboratory equipment they worked with, or even use new equipment purchased by or on loan from their corporate sponsor. Teachers tend to integrate science and mathematics more in response to the lack of "subject boundaries" in applied industry work.

Teachers also take a different view of problem-solving. Many teachers have noted that workplace problems are more challenging and complex than the contrived problems they use for demonstration purposes in the classroom. In response, most IISME fellows devise strategies for modifying assignments or class projects to provide more challenging, open-ended problems for students to tackle. As one example, a mathematics teacher assigned groups of students the task of planning a public park and presenting their plans, including economic analysis and graphic layout, to a mock city council.

As one of the oldest and largest programs of its kind, IISME has been instrumental in helping launch other IISME-like programs throughout the United States and abroad. Summer industry fellowship programs for teachers now exist in over half of the United States and in Denmark. Many other programs are in the formative stages. Contact IISME, c/o Deskin Research Group, 2270 Agnew Road, Santa Clara, CA 95054 for a list of state contacts.

The success of IISME has resulted from bringing two disparate cultures together, allowing industry and education to learn from each other and to gain respect for each other's capabilities and needs. Through programs like IISME, industry leaders have gained awareness of the challenges facing schools, formed strong ties to educators, and developed a deep commitment to the education of our youth. As a result, many have joined important education initiatives — not to tell schools what they are doing wrong, but to participate as equal partners in the effort to improve mathematics and science education in this country. Schools and educators have benefited by utilizing industry ideas and resources in the classrooms, as well as by gaining a better understanding of the career marketplace their students will enter after graduation.

Training the Scientific Community

BY HECTOR TIMOURIAN

In our highly technological society, scientists, engineers and technicians have the responsibility to do more than good science. The scientific community (including academia, private industry and government research laboratories) needs to become involved in science education. This involvement should be a partnership at all levels from kindergarten through graduate school.

If we agree that scientists need to become partners in precollege education, what are the expectations of this involvement? Do we send scientists into the classrooms to give lectures to students? Do we expect scientists to rewrite curricula and textbooks? Do we have them train science teachers? These tasks use scientists as resources for accurate content knowledge. Without detracting from what many scientists are already doing, I suggest that they are in the unique position of sharing the enthusiasm for discovery with teachers and students. Scientists can promote careers in science and technology and help students understand what scientists do. When scientists visit classrooms, students and teachers get to meet and to know scientists as "normal" people.

The Lawrence Livermore National Laboratory Science Education Center (SEC) has developed a workshop to help scientists become involved in education. A major outcome of the workshop is having scientists learn how to share the joy of discovery with teachers and students. In one technique the scientists recall why they went into a science/technology profession. This is done by having them fill out a questionnaire that is then used to facilitate group discussions.

Many of the scientists recall having a special place or "lab" where they started their own investigations. Some started taking apart clocks and radios just to find out what made them work. Another scientist remembered watching ants on her windowsill and experimenting with setting up a variety of obstacles. Another scientist recalled putting a pile of dominoes under the middle of a wooden ruler and then placing objects on the end of the ruler to compare their weights. For some, it was the help of a special teacher or becoming involved in a science project. Women scientists recall how their skills in math and science helped them overcome cultural expectations and realize that they wanted to become a scientist.

Scientists can convey their interest and enthusiasm directly to students. One geologist explained that rocks fascinated him when he was a child so he began a rock collection. This led him to wonder why the rocks were so different. When he visits classrooms, his fascination with the subject is contagious. Students become interested in examining rocks, making observations, and learning that they can infer a history of how a particular rock got its shape and structure.

In addition to helping scientists articulate the motivational factors that ignited their interest in science, the SEC workshop aims to have scientists a) become aware of education needs, b) become knowledgeable about student competencies, and c) begin to develop a presentation or lesson that brings something of their work into a classroom. This workshop was itself developed with the help of teachers and school administrators.

Many scientists and educators place too much emphasis on factual learning. Discussions during the workshop help them realize that it is more important to appreciate conceptual knowledge instead of remembering facts. Scientists become aware that their knowledge of facts is not as needed as their help with the formulation of conceptual frameworks that connect several ideas. This discussion also introduces scientists to the latest developments in science education, such as Project 2061, the NSTA's Scope and Sequence, or the California State Science Framework.

These discussions of major ideas in science are expanded by having the scientists list the major concepts that they think students should be learning in school. A few concepts are then analyzed to determine for each:

What information is needed to explain the concept?

How do we obtain this information?

What tools are used in making observations?

What is still unknown about the concept and how can that be explored via further experimentation?

How can students relate to this concept?

Is there a technical application that directly affects them?

Where does this concept connect with other disciplines and concepts?

Through this analysis, scientists help fulfill a major educational need, showing with specific examples how our scientific knowledge integrates observations, inferences and experimentation in the formulation of concepts.

The workshop helps scientists realize that children are not little adults and also helps connect them with the big kid within themselves.

In another part of the workshop, scientists are introduced to background knowledge that reviews pedagogical literature describing how children learn at different ages (as pioneered by Piaget). The scientists also learn about instructional strategies that best conform with the learning abilities of students of different ages (see Table 1). One thing that is very clear from listening to educators is that any interaction with students must take into account the students' capacity for learning, including such factors as eye/hand coordination and attention span. Many a scientist has visited a classroom and presented accurate science at a level that was far beyond the students' comprehension.

This theoretical knowledge is brought home by helping scientists remember how they felt as children. We engage them in a number of simple problem-solving activities. In one of these activities, we give them two toothpicks and two drops of colored water on a piece of wax paper. They need to make observations and determine differences between the two drops. This type of discovery activity helps the scien-

---

tist realize how the excitement and concepts of science can be shared with others using very simple materials and ideas.

The workshop helps scientists realize that children are not little adults and also helps connect them with the big kid within themselves. After learning about the capabilities of students at different grade levels, scientists can prepare to make the most effective use of the time they spend in the classroom or in helping teachers to prepare lessons.

Even if they never go into a classroom, scientists need to learn how to communicate what they are doing. An important part of the workshop is to have the scientists begin to develop a lesson that brings something of their work into a classroom. In doing this, scientists take from the previous discussion a major concept that they can present in the context of their work. Working with other workshop participants, the scientists can develop the plans for their lesson, including demonstrations or hands-on activities that illustrate the central concept. For example, a scientist whose work involves sorting chromosomes can focus on the sorting and classification process. Many hands-on lessons can be devised using simple materials such as leaves to explore sorting and classifying at different grade levels.

A vital aspect of the partnership involves the ability of the scientist to get feedback from teachers. The scientist needs to get feedback and this can be as difficult for the scientist as it is for the educator. When a scientist goes into a classroom and is not intelligible, the teacher is often reluctant to criticize, not wanting to hurt the scientist's feelings or appear to be ignorant. There is no easy way to promote feedback, but this is a crucial element in the partnership.

We have found that a longer period of time than a two-day workshop is needed to develop a relationship that enables honest feedback and that results in lessons that are effective in the classroom. We accomplish these objectives in the context of summer institutes that bring scientists and teachers together for one to four weeks. During the summer institutes, teachers and scientists work together with a focus on developing lessons for the classroom. Teachers value the experience of working with scientists and they also appreciate the activities that they can bring directly to the classroom. The institute format enables the scientists and teachers to explore the concepts in as much detail as the teachers

TABLE 1

Suggested instructional strategies that best conform with the learning abilities of students at different age levels.

	Primary (K-2)	Elementary (3-5)	Middle Grades (6-8)
Learning Abilities Students at these grade levels:	• believe only what they see, e.g., change in shape means change in quantity • learn through manipulating objects • concentrate on parts, not the whole • can pay attention for 10-15 minutes	• begin to generalize • begin to understand concepts • can believe without seeing • like to memorize facts • can pay attention for 20-30 minutes • can follow a task for several days	• can hypothesize • can conceptualize abstractions • begin to understand multiple causation • can challenge authority • can handle 30-40 minutes, and can follow multiple tasks
Teaching Strategies Students are best taught by:	• making observations • simple manipulations • pictorial communications • simple comparisons	• building relationships • using space-time relationships • formulating inferences • drawing simple conclusions	• formulating experiments to test hypothesis • recognizing and predicting patterns • developing models to explain

Adapted from California State Science Framework and *Sharing Science with Children*, a North Carolina Museum of Life and Science Publication.

need for their own understanding and for tailoring the information to their students' level. It is during these give-and-take sessions that a trust is established and the scientists become more open to feedback to optimize their lessons or presentations to best meet the needs of students.

By working together in developing a lesson, the scientists provide the technical content and the teachers provide the knowledge of how to transmit the information to the students. By engaging in this practical task, the scientists become aware of how to communicate their knowledge to those who do not have technical background without feeling that their science content is being corrupted. Once the scientists have been able to communicate their knowledge to the teacher and develop lessons, they then are excited to determine if they can communicate with students. An important aspect of this communication is simply sharing their excitement about science. When all goes well, students get interested in science, learn important concepts through hands-on discovery processes, and experience scientists as real human beings who do interesting and relevant work.

In *Hard Times*, Charles Dickens points out the analogy between education and the industrial revolution's assembly lines. Students are equated with little pitchers on an assembly line to be filled with facts. In that context, education reform involves tinkering with the facts (textbooks), the assembly line workers (teachers), and the factory foremen (principals and other administrators). However, we are beginning to learn that true education reform is much more than adjusting the assembly line. For systemic educational reform to happen, students must be empowered to become active explorers, not empty vessels waiting to be filled. Scientists, by the very nature of their profession, are explorers. By working in partnerships with teachers, scientists can help ignite curiosity and excitement as well as teach the science process skills that channel that curiosity into productive learning experiences.

University Research
Expeditions Program

BY JEAN COLVIN

For 15 years, the University Research Expeditions Program (UREP) has been inviting the public to join the scientific community in exploring the wonders of planet Earth and in searching for solutions to the complex issues facing its survival. From modest beginnings in 1976, our program was built upon the firm conviction that curiosity about the world is not limited to scientists in lab coats or academics secluded in ivory towers. We knew that men and women of all ages, from all walks of life, would eagerly accept the opportunity to share and support the challenge that underlies all research — the need to understand.

When I helped launch UREP in 1976, the idea of having members of the public participate in university research was so unusual that no one in the university system quite knew where to place UREP administratively. The director of the University Herbarium was so supportive of the idea that he provided a desk, an antiquated typewriter and a phone in a cubicle where 200-year-old botanical journals were stored. The program remains under the UC Berkeley College of Letters and Science even though it has grown to sponsor research from many departments and all University of California campuses. Hundreds of field research teams have participated in over 50 countries worldwide.

UREP provides field research support on important, interesting issues in environmental studies, biological and health sciences, the social sciences, and the arts and humanities. Scientists from the University of California lead the teams, which engage in a wide variety of activities such as counting, collecting and cataloging living specimens; investigating the competition between native and alien grasses and assessing the role of fire; testing sustainable technology using renewable energy sources to meet daily living needs in a Third World village; and observing and recording monkey behavior in an African rain forest. A typical team consists of the UC field research director, two to four teachers, one or two undergraduate students, and several donor-participants from the general public UREP program.

Teachers participated in the field projects from the very beginning. However, we did not begin a formal Teacher Research Participation Program until 1979 when UREP received a grant from the National Science Foundation (NSF). I had seen an NSF announcement one week before the deadline. I called and spoke with a program officer who became very excited about the idea of teachers participating in research and encouraged me to try to meet the deadline. I had never written a grant, but, with good advice from Professor Carl Helmholz of the Physics Department and Bob Rice and Ted Beck at the Lawrence Hall of Science, I made the deadline and the program was funded. The first program of its kind in the nation, UREP's Teacher Research Participation Program has provided hands-on research experience to over 500 K-12 teachers who return to their classrooms with a better understanding of the scientific process and of the cultures and environments in which the expeditions take place.

> *Convincing scientists that "nonacademics" could assist in their research was no easy task. Part of the UREP goal was to break traditional barriers and to create new avenues of communication between scientists and precollege teachers.*

Initially, the program drew teachers from all over the country, but it soon became clear that part of the value of the program was the interaction of the teachers before and after the actual field activities as well as the opportunity for on-going interaction with the university. As a result, pre- and post-expedition orientations and curriculum development workshops were added to the program and recruitment was focused primarily in California so teachers could attend the on-campus workshops and work together on the curriculum development.

Convincing scientists that "nonacademics" could assist in their research was no easy task. Part of the UREP goal was to break traditional barriers and to create new avenues of communication between scientists and precollege teachers. Once a few researchers had successfully taken the step, others became interested. Because many researchers had not been in contact with precollege teachers since they were in school themselves, we instituted a special workshop designed to facilitate the interactions. To make sure that the faculty would attend, we billed the workshop more as a focus on topics like "emergency procedures" and "accounting in the field" rather than communication skills. Once they were there, most researchers agreed that the communication activities and group problem-solving exercises were very worthwhile in preparing them to work with their "nonacademic" team members.

One of the activities is called "Realistic Expectations." The team leaders and the team participants meet as two separate groups. Each group makes a list of its expectations for the project and then compares the lists. The entire group then discusses common objectives and what expectations are realistic. Leaders tend to want their participants to be as knowledgeable and interested in the research as they are. In preparing the field activities and schedule, they need to realize that teacher participants have a varying and wide range of experiences and interests.

In the "Scenario" activity, a group of 6-10 leaders and participants confront a potential problem situation. In one scenario, project leader, staff and most of the participants have developed a good working relationship during a botanical expedition high in the Sierra Mountains. All seem content except Mr. Carp, who is never satisfied: "The food is lousy, the weather is crummy, the accommodations are poor, the work is not fairly distributed, etc." Mr. Carp also does not like the way the research is being conducted and insists his way would be much better. As a result of his constant complaining and bids for attention, the project leader spends a lot of time with him to the detriment of the rest of the group. What was once a highly motivated team becomes a loose gathering of disgruntled individuals. The workshop group determines the best solutions to the situation, how it could have been avoided and the different responsibilities of participants and leaders.

The spring orientation workshop does more than enhance interpersonal relationships. This workshop combines slide/lecture presentations by

---

UC scientists on the field research projects with presentations of curriculum units developed by previous teacher participants. In addition, teachers receive project packets that were developed by the staff and field directors. These packets provide the background for the field research; tools and methods to be used in the field; considerations of foreign customs, climate, logistics; and reference materials for further study and preparation. Working individually and in small grade-level groups, each teacher develops a preliminary plan for a curriculum unit based on their research project. These are formalized with the teacher's principal as a Team Plan prior to departure for the field.

After the field experiences, which generally last two or three weeks, teachers attend follow-up workshops designed to provide them with skills and ideas to develop their own curricula. Workshop presentations include examples of hands-on science activities, brainstorming sessions, exhibit preparation, publishing articles for teaching magazines and journals, conducting student field research projects, and tips on how to conduct an in-service workshop. Teachers also have the opportunity to share information and materials that they have collected and are able to collaborate on further ideas for lesson plans.

To disseminate the materials and activities beyond the individual teacher, each participant is obliged to present the curriculum module at 1) a symposium for all teacher participants, 2) a school or district in-service and 3) a community or professional meeting. Three UREP teachers recently attended the annual National Science Teachers Association meeting to present curricula they developed as a result of their participation. Copies of all the curriculum projects developed by the teachers are available through UREP's Research Preference Library.

An important goal of this program is to stimulate greater understanding of and interest in science careers among underrepresented minorities. Many of the field projects are located in developing nations in Africa and Latin America, and most include host participation. Given the increasing ethnic and language diversity of today's student body, links to these regions are of particular benefit to help teachers develop science activities that will interest these students. These cross-cultural experiences also provide an excellent opportunity for U.S. educators to enhance their language skills as well as their sensitivity and abilities to teach an ethnically diverse student population.

As one example, Spanish-speaking teachers can participate in UREP's recently developed Workshops in the Forest program in Ecuador. This project involves teachers and scientists from Ecuador as well as the U.S. teachers and scientists. The workshops offer participants a unique opportunity to share information and resources in an exchange that fosters a global, interdisciplinary and culturally sensitive approach to environmental issues. Participants work together on site at biological field stations in different regions of Ecuador to develop bilingual environmental educational materials and activities based on the group's own field experience (see accompanying story).

UREP's experience has shown that the intensity and informality of the field experience breaks traditional barriers and allows for a much greater exchange and interaction than normally occurs in professional development courses or laboratory situations. The opportunity to work on a scientific expedition, often in unusual and remote locales, attracts teachers who might not otherwise participate in a professional development program. Teachers gain a better understanding of other cultures and their scientific and social needs and priorities.

Almost 75% of the teachers report on-going contact with field directors and with other teachers in the program. Some have arranged classroom visits by field directors, or have taken student field trips to local research sites and facilities. For example, one teacher took her class to a local archaeological site where the students participated in the excavation efforts in collaboration with one of the field directors; another teacher took a group of students to a UC animal behavior lab for a tour of the facility and presentation by the director who had conducted a study of penguin behavior.

Almost all the teachers responding to the follow-up survey report that they would participate in another scientific field experience citing the improvements it made in their curriculum, in their enthusiasm for teaching, on the favorable reactions of students towards scientific research, and because it stimulated them to seek out local resources and learn how to use new tools and equipment. Among the other insights gleaned about the program from evaluations were the following:

---

1. Curriculum development is a time-consuming process. It often takes more than a year to develop curriculum fully and assess the implications of a scientific field research experience;

2. Teachers respond extremely well to peer teaching, collaborative curriculum development opportunities, and contact with scholars in their fields; and

3. As a result of working as a scientist and making a contribution, the image of teachers improved among both students and administrators.

Perhaps the most fitting conclusion to this article is to describe a sample offering from this year's catalog. Entitled "Alien Plant Invasion of Hawaii's Woodlands," this project in Hawaii Volcanoes National Park investigates the invasion of the big island of Hawaii's dry tropical forests by a variety of non-native grass species. These grasses were introduced to Hawaii as forage for cattle and have since invaded native woodlands where they promote fires by creating a continuous layer of dry litter in a forest otherwise lacking such fuels. These fires in turn appear to have a negative effect on the survival of native species. The project will identify why alien grasses are more successful than native grasses at occupying these habitats and study the influence of fire (and the post-fire presence of alien grasses) on nitrogen cycling in these nitrogen-starved forests.

Under the direction of Dr. Carla D'Antonia of the Integrative Biology Department at UC Berkeley, team members will live and work at 3,800 feet on the slope of Kilauea Volcano. Participants will prepare plots of land for a "controlled burn" by removing and sorting the vegetation. Core samples will be taken before and after the burn to measure soil nutrients and to assess the number and variety of seeds present. Participants will also help with vegetation mapping. Increases in seismic activity give ample warning as to upcoming eruptions. Explosive eruptions are unlikely during the expedition!

ONE TEACHER'S STORY

From my classroom in Oakland Junior High School to a mountain rain forest in Ecuador was a long trip, but a very rewarding one. Under the auspices of University Research Expeditions Program (UREP), five other California teachers and I traveled to Maquipucuna Reserve, about 70 miles north of Quito. This reserve, only two years old, was founded by a group of Ecuadorians concerned with rain forest destruction. With funds provided by the Nature Conservancy, the reserve recently expanded and now protects ten square miles of rain and cloud forest. This patch is part of a much larger range of forests cloaking the slopes of the Andes. Under constant encroachment, this forest is now only 9% of what it once was. The reserve shelters beautiful orchids pollinated by nectar-eating bats. Bromeliads burden the trees, and climbing vines wrap tree trunks. Tree ferns stretch 40 feet into the air, their trunks bristling with thorns, while leaf-cutter ants ply the trails with their moving mosaic of green leaves. The reserve is virtually unknown to many of the villagers who live within a few miles of it.

Working with Professor Grady Webster of the University of California, Davis, we collected plant specimens to be preserved and sent to herbaria in Ecuador and the United States. Then, in a special workshop supported by a grant from the National Science Foundation, we worked with five Ecuadorian teachers from the elementary and secondary schools of the nearby villages on curriculum devoted to the rain forest.

Working primarily in Spanish, we exchanged ideas and concentrated on those which Ecuadorian teachers thought would be practical. Some ideas tested techniques for conservation and fertilization. Others stressed the importance of the great diversity of life in the rain forest. The adaptive value of different leaf shapes was another lesson, as was the complex water cycle of the forest.

On the last day of the workshop, we got a chance to share our new lessons with the local students. We piled into our vehicles and drove to Nanegal, a village about 15 kilometers from the reserve. The entire student body of Colegio Tecnico de Nanegal (about 100 students, grades 7 to 12) assembled to participate in our lessons. The room came alive as plant samples were passed around, maps of global rain forest destruction were displayed, and biodiversity was discussed.

We returned to California determined to teach our students about the forest, and what they might do to protect it. But we left behind teachers with an even greater challenge; to teach the youth of the villages in the forest to coexist with the forest, to live with it without allowing it to be destroyed. For ultimately, the forest depends on them most of all.

This report by Anthony Cody appeared in *California Classroom Science,* a publication of the California Science Teachers Association.

A SCHOOL OF EDUCATION EXTENSION SERVICE MODEL

BY PAM CASTORI

"**Y**es, I have left the classroom." This is the response that I have given to teachers these past two years because of a response that I made to a flyer that I received in the mail. The flyer advertised for a Science Education Extension specialist whose responsibilities would involve working with educational and science researchers at the University of California at Davis in collaborative research and service activities with schools. My teacher colleagues wonder how I could desert my students. What could I possibly do that could have as much impact as I had by teaching biology and chemistry in high school and working with elementary schools in my district?

Well, the job looked like it offered many opportunities for continued learning and teaching about science and science education. In fact, that is exactly what it entails. The position is with the Cooperative Research and Extension Services for Schools (CRESS) Center at UC Davis. The CRESS Center is administered by the Division of Education at UC Davis to support research and development activities that have the potential to improve K-12 schooling. The CRESS position offered the opportunity to teach and learn in a different context — one that combined university education research with classroom teaching practices.

What would a Science Education Extension specialist do at the CRESS Center? Part of the answer lies in geography. UC Davis is located in Central California, world renowned for growing food. In fact, UC Davis has a very prominent School of Agriculture. That practical academic domain features an Agricultural Extension Service whose primary purpose is to connect the academic world of research with the practical world of growing food and plant products. An Agricultural Extension specialist is the personalized connection between the practitioner in the field and the researcher in academia. The connections go in both directions. Some research questions grow out of problems encountered on the farms. In the other direction, people who work in the field obtain valuable information and services from the university researchers.

Why is this a good model for connecting a Division of Education and the precollege schools? Ask any teacher how much they apply in their classrooms educational research that is conducted at the nearest university. The odds are very high that you will get a blank stare or strongly worded comments, perhaps unfit for students' ears, that education research is disconnected from the real world and concerns of the classroom teacher. In contrast, the model of the Extension Service and the Extension specialist attempts to create meaningful connections by bringing the worlds of educational theory and educational practice together in nonthreatening and useful ways. As in the case of agricultural extension, the university aims to help answer questions and provide useful services to the site practitioners, namely K-12 teachers, administrators, parents and students. In the process, the field helps identify important questions for research and actively participates in designing and conducting that research. Instead of the school and its occupants being passive objects of study, they collaborate in the research process.

As a teacher who is always seeking ways to improve my practice, the position of Science Education Extension specialist promised to provide me with unique experiences and insights. The challenge involved in the specialist job position is to connect K-12 school people with interested educational and scientific researchers to explore common questions and concerns, to learn from each other, and to use this partnership model to improve the science learning experiences of children in the schools.

The CRESS center sponsors many activities. Rather than present an exhaustive list, I will focus on several programs that illustrate key features. The center is the home for the area's Subject Matter Projects. These professional development projects for teachers evolved from a program called the Bay Area Writing Project that proved so successful that it has transcended geographic and academic discipline limitations. In California, in addition to the numerous sites of the Writing Project, there are now also many sites for the Mathematics Project, the Science Project, the History and Cultures Project, and the Foreign Language Project.

What is a Subject Matter Project? Variations exist but there is a common philosophy to these staff development programs for teachers. The features of the Science Project below would apply with some variation to the other subject matter projects. The Science Project assumes that:

1) "The best teachers of teachers are other teachers." Teachers have enormous amounts to contribute to improving science education and to the training of teachers.

2) Teachers need to experience what the scientist (or writer, mathematician, etc.) does. The teacher engages in activities that replicate what a scientist does. This experience often happens in summer workshops with considerable input from scientists.

3) The experiences and activities for teacher participants reflect and model the constructivist educational approach. Teachers have time and situations in which to reflect and construct their own meaning of science and science education. They re-evaluate their assumptions about the subject and the teaching of the subject.

4) A learning community develops and evolves. Teachers become part of this community and may participate for years.

A unique feature of the CRESS Center is that it is the center for many of the region's Subject Matter Projects. Obviously, this feature facilitates interactions compared to an area where the Science Project is headquartered 80 miles from the Math Project. Proximity provides a context for the exploration of inter-disciplinary and cross-disciplinary concerns. It has also sparked collaborative education research projects such as the one described below.

Item 3 above refers to the constructivist educational approach. The philosophical basis of constructivism is that I construct my own meaning. No one can make me understand something. I am not an empty vessel that gets filled. My mind is not a chalkboard that has erroneous information that can be erased and then corrected by having someone write the correct information on it. Instead, all of us have to construct our own meanings. We are responsible for our learning. In brief, constructivism is an educational approach that:

1) Accesses a student's prior knowledge and conceptions. Students become aware of the knowledge, ideas and models that they already have, often without previously being conscious that they had this information or concepts.

2) Uses the context of the student's conceptions and knowledge to design experiences that will lead to the teacher's goal. There is an overall structure but it is flexible to enable students to have different starting points and paths that lead them to new understandings.

3) Challenges students to apply their new understandings to a new situation and see how it fits.

4) Helps students become aware of their ideas, their thinking processes, and how their ideas may change.

The CRESS Center sponsors and facilitates collaborative interactions among teachers, members of the UC Davis Division of Education, and faculty in different disciplines at UC Davis. For example, through CRESS, I became very involved in a collaboration that examined connections among the Subject Matter Projects and the constructivist approach. Three teachers, three CRESS staff members, and a physics professor formed a Project Design Team to examine issues of common interest. We brainstormed and defined our project questions. Two of the final questions were:

1) Do changes in teaching strategies in one discipline (as a result of summer institutes) carry over to other discipline areas? Does participation in a Writing Project, for example, affect the way the teacher presents science or math?

2) Do strands reflecting the notions of constructivism exist across the disciplines?

The Project Design Team began the long task of designing how to collect data from teachers who had participated in the Subject Matter Projects. The teachers on the Design Team had a major role in defining

---

SCIENCE EDUCATION PARTNERSHIPS

the scope of the project and its methodology. The team developed interview and observation criteria after many hours of hashing out a working description of constructivism; how the institutes reflect constructivism with or without explicitly using the term; what features of the classroom might be used as indicators; and how to collect data and conduct interviews without intimidating teachers and students. We analyzed the data from the classroom visitations and from whole group meetings and reached the following conclusions with respect to the questions described above and the sample of teachers who participated in the study:

- Teachers who participate in Subject Matter Projects incorporate elements of constructivist teaching strategies in the subject matter for which they attended a summer institute. The elements observed most often were student-centered activities involving cooperative group work and open-ended assignments/problems.

- Teachers do employ strategies to assess prior student knowledge. However, the teachers rarely use that information in deciding what lessons they will present or in altering their instructional plan.

- Despite expressing enthusiastic interest in the concept, teachers have difficulty transferring the understanding they gained in their particular subject area institute to other subject areas.

- Teachers who do transfer understanding/strategies from one subject to another are more likely to teach in ways that reflect constructivist notions of learning.

- Teachers feel empowered by learning about constructivism. It gives them a theory to express what they feel is true about the way their students learn and what kinds of conditions are necessary for conceptual understanding to occur. However, explicitly presenting constructivism as a named concept does not ensure that teachers will teach in ways that reflect constructivist notions of learning.

This project has produced several practical results. Three of the Project Design Team members (two of them classroom teachers) taught a course in the UC Davis multisubject credential program. Titled "Bridging the Gap," the course used a constructivist orientation to link theory and classroom practice, link the classroom teacher with the university, and to create connections among the ideas and approaches of the different subject matter projects. It represented the first time that classroom teachers had primary responsibility for a course offered for credit in the UC Davis Division of Education.

The study also highlighted a very real obstacle that requires further research and development for implementing the constructivist approach in the classroom. While it is comparatively easy for teachers to access prior student knowledge, it is much more difficult to tailor lessons that depart from that variable base and lead along different paths to the desired learning outcome. The teacher needs to have a variety of explorations available for students that will help them construct a converging concept from different starting points.

The example that I chose illustrates how classroom teachers can work as collaborative partners with a university Division of Education. Other current CRESS research projects focus on the academic achievement of linguistic minority students; challenges and opportunities faced by teachers in restructuring high schools; institutional and professional incentives for encouraging teacher change; and student response to different strategies for teaching science, mathematics, writing and Spanish. Elementary and secondary school teachers conduct systematic research inquiries about teaching and learning in their own classrooms. Through an active program of publications, colloquia and workshops, the CRESS Center enables university and K-12 educators to share and refine their understanding of schooling and the challenges of planned school change.

As a result of my CRESS experience, I have joined the small but growing ranks of science educators who speak the languages of both the classroom teacher and the university education researcher. In discussing science education partnerships, Art Sussman talks about "transfer RNA people" who know the worlds of both scientific research and precollege education. As we continue in our attempts to systemically reform science education, it has become clear that we need many people who can bridge the different worlds that can significantly contribute to this reform process. The CRESS program at UC Davis provides an institutional framework for creating and nurturing those vital connections.

PUTTING THE FOCUS ON PRESERVICE SCIENCE EDUCATION

BY GLENN CROSBY

*This article is adapted from a speech given at a conference in
San Diego of the American Association for Higher Education.*

We have very serious problems in this country in education,
particularly in science education, and some of the problems stem from
poor teacher preparation. From the standpoint of improving the
schools and improving what teachers do, I think there is a misplaced
emphasis on inservice versus preservice education of teachers. For those
unfamiliar with the terms, inservice refers to educational experiences
that are provided to teachers who are already certified, whereas preservice is the training that is provided for prospective teachers. I do not
think permanent solutions to the nation's educational problems are
going to come from a focus primarily on inservice education.

The educational programs for potential secondary and elementary
teachers have never been holistically organized by universities. We simply do not have cooperation among exponents of the subject disciplines
and exponents of pedagogical methods. We do not have that interaction, and the prospective teacher suffers as do all his or her eventual
students. I would like to see a sterling example of preservice education
that integrates science content and process with the pedagogical skills
needed in today's schools. I have never observed it.

Part of the problem is the lack of a feedback loop to adjust preservice
education for what actually goes on in the schools. We have very little

information coming back to the universities regarding what teachers really do. We in higher education are being asked to educate people who go out into the field and we do not understand or take into account what happens to them when they leave the university. To make a mundane analogy, would a car manufacturer sell automobiles without ever finding out how the cars are performing, what problems are occurring, and how these problems could be prevented?

> **The way most universities are currently constituted, they are not suited to the task of educating teachers at any level.**

A related part of the problem regards inservice. I have been involved with many inservice programs since 1986, and I have concluded that inservice is misused. Much of it is remedial, and that is not what it was designed for. Teacher inservice was designed to be progressive, to introduce practitioners to the cutting edge, and I find instead that it is really remedial education. So, in my opinion, there is something basically wrong with inservice programs as they are practiced.

Another situation that I have recognized is that the modern university — with its specialization and its departmental walls and its College of Education and College of Arts and Sciences — is not properly set up to attack the problems of teacher education. The way most universities are currently constituted, they are not suited to the task of educating teachers at any level. This is true for all prospective secondary teachers but it is particularly true for those planning to teach in the sciences and certainly for prospective elementary teachers.

After many, many years of teaching I have also concluded that the key to any reform of education is the teacher. So, I am concerned about all this business about standards and the application of standards. I am very much concerned about enunciating educational goals by the year 2000, without taking into account the people who are supposed to power the schools to reach these goals, namely the teachers.

If we are going to do anything to improve science education in the public schools, then we must involve the universities in a major way. And they are going to have to start focusing attention on preservice programs and shift the focus away from inservice education. Now, I am not against inservice education, but it is too easy to sell. Everybody loves inservice. The teachers love it because many of them get stipends. The professors love it because they are paid summer salaries. The districts love it because it brings in money. Congress loves it because it funnels money directly to constituents. In contrast, there are few incentives to improve preservice programs for prospective teachers, and consequently there are few champions in that area.

Another condition that will impede educational reform via inservice programs is that many of the teachers whom we need to reach do not attend inservice programs. We have all seen the same enthusiastic, committed teachers attending the conferences, workshops and planning committees. In preservice courses, I have a captive audience. I award grades and I can demand performance. With an inservice program, I must cajole and coax. I must do all these things in order to get the teacher there, particularly the unprepared teacher. Everything considered, I think much more emphasis should be placed on preservice, rather than inservice, to power systemic reform.

One can see the nature of the problem in another way. Teacher turnover is very large and each year thousands of new teachers are graduated who, the day they graduate, are ready for inservice training. They have just finished being educated for their jobs but they do not possess the techniques, the methods, and the knowledge commensurate with the problems they must instantly face in the fall. I know this for a fact because I have run institutes for teachers funded by the National Science Foundation. One was a leadership institute, and the rules said we were allowed to accept only teachers who had been out in the field at least three or four years. Actually, that is a good idea because they have some inkling of what their problems are. Yet, I received many applications from people who had just graduated. They called me up saying, "I've just been in the school for one year, and I realize I do not know what I need to know in order to teach." And, I thought, "This person just came out of my institution with a degree as a certified teacher, and now what this person is telling me is that the whole program that we have for that teacher is useless."

I realize that politically it not going to be easy to emphasize preservice instead of inservice. To achieve a renewed emphasis on preservice, we need to look holistically within the university at the education of teachers. I am a chemist and the American Chemical Society (ACS) has an approval program for institutions that offer a bachelor of science degree in chemistry. That degree, the bachelor of science in chemistry as approved by the American Chemical Society, is not the best degree for someone who plans to teach chemistry in high school. The reason is — it is not broad enough. It focuses too much on producing someone who is going to go either into industry or into graduate school. It is not appropriate preparation for a precollege teacher's challenges and necessary knowledge base.

Virtually all colleges and universities in the United States aspire to award a degree in chemistry that is approved by the American Chemical Society. So you can see the problem we have for those who wish to go into the teaching profession. The ACS is making some headway toward changing requirements and has actually approved an option: Chemistry/Education. I think it is significant, however, that, as far as I know, only one school in the country has such an approved program, even though the option has been in existence for a couple years. I think that shows the lack of both understanding and concern by my colleagues in the chemistry profession concerning the plight of prospective high school chemistry teachers.

As I said before, we need to establish feedback loops. As a university chemistry professor, I need to know what is happening in the secondary and elementary schools, but we do not even have communication between colleges within the same university; we do not have feedback between the College of Education and the College of Sciences; we do not have feedback between the mathematics department and the chemistry department; we do not have physicists and chemists talking to one another. We do not have this basic information exchange occurring in the modern university.

Once that modus operandi worked well for producing specialists. Now it is not even working well for that objective, and we are creating hybrid programs at the graduate level. First one had biochemistry, and then chemical physics, and now we have material sciences. At the graduate level the fields are joining and the boundaries are blurring. But, for teacher education and many undergraduate programs, that fusion has not happened yet. Most of the people who are graduated from college really possess no purview beyond a very narrow specialty.

Let us turn to some specifics. A number of years ago, I decided to run an inservice program for middle school teachers. This was my first real jump into middle school. I am an experimentalist and I think the best way to learn anything is to give it a try. I came away from that month shaken. I was appalled. What appalled me was the fact that these people desperately wanted to teach science in their classes; yet, their basic knowledge was so poor that I really had difficulty helping them.

It was not the fault of the teachers. They had come through the normal programs. So I examined what prospective teachers at my institution actually took in science. And I found out that most of them, 90% or more, took one course — an introduction to biological sciences — and that was it! In other words, their basic education in science was virtually nonexistent. These people wanted to teach some science to kids and to excite kids about science. I thought that is an impossible task for them. So I said, "Why don't we put together a coalition at the university? Let us improve the quality and quantity of science that prospective elementary teachers take."

The obvious solution would be to have them take a year of chemistry, a year of physics, and a year of biology. But if one looks at the programs that we have in chemistry, physics and biology, those courses are not the ones those people should be taking. More than just impractical, those abstract, sophisticated and highly mathematical courses would create a fear of science in the prospective teacher.

So we wrote a proposal to the National Science Foundation to develop some science courses for elementary education majors. And, this is where I began to learn, after being in the university for 30 years, that it is not easy to interact with the Education School. And it is not easy to interact with your colleagues in the physics department on educational matters. It is not an easy job, but you can find one or two people. Maybe you can find a dean who will listen. We needed a commitment of this kind, "If we produce something that is useful, will you require it?" That was the condition. I have learned in my old age that there is no use doing things unless they are going to be institutionalized. If they are not going to be institutionalized, do not do them because you are just wasting your effort.

We put together a small team and wrote a proposal to the National Science Foundation to produce two new courses. I began to work on a course that is now half chemistry and half earth science. The earth science part is taught in the geology department and the chemistry part is taught in the chemistry department, with labs. And other people developed a course in astronomy and physics. We worked three years to develop those courses. Finally, we gingerly offered them on a voluntary basis. Remember, the condition was that if we could make them work, these courses would be required. Well, it became very interesting to me to find out that this could catapult my university into the top 1% in the nation with regard to requiring courses in science for elementary education majors.

We started offering the new sequences for a small number of people, nine or 10. I taught the chemistry course. The laboratory stumped me. I did not know what to do. What would one do? One would pick the simplest of the traditional lab experiments. I taught the course with a lab and it was a disaster. Total disaster. I gave myself a "D+" when I finished. Why? Because it became clear to me that I had not analyzed the situation these people were going to face. So, it was an irrelevant laboratory, and, to a large extent, an irrelevant course.

To attack educational problems we must get the scientists to do what scientists generally do — look at the data, and figure out from the data what is really needed. I went out, talked with elementary teachers, and I looked in the schools. I saw that there was no sink, no money for supplies, and no stockroom. I asked myself, "What kind of laboratory must I develop that will give these teachers something they can use, something they can be comfortable with in their classrooms?"

I suddenly realized the principle I could use. I decided to run the entire laboratory based on consumer products and things that you throw away — pop bottles, beer cans, and so forth. The supermarket and the drugstore would be the chemical supply house. It can be done, and it is relevant. Moreover, it is a lot of fun. And the prospective teachers now love the lab. The lab is interesting to them because it uses the things they know about. The lesson I learned was that you must throw away all the things you have been doing and start anew. Well, the course has now gone through two iterations. I have gone through some iterations, too. I realized that if I wanted these teachers to teach hands-on science, then I had better teach them with hands-on science. So I began to change my "lectures." By now I had 80 people in the class, and it was required. Not without a great deal of fight, by the way. But it is now a required course for elementary education majors.

If scientists are going to become involved in this business, they must realize that they have a lot to learn. There are people who know something about pedagogy. They are in the Education School. The problem is — how do we interact with those people? I have not solved that problem, but I think that the only way is to contact a few souls who are interested. Unfortunately, there do not appear to be very many who are really interested in helping the kids. We are interested in publishing papers; we are interested in doing lots of things. There are not that many people in universities who really care about helping kids. And so, you have to find that very few. And then, you have to support them and build the new structures around them.

We have built an administrative structure in our university that we call "SMEEC," a terrible acronym for Science, Mathematics and Engineering Education Center. I call it neutral ground — it is where people in the Education School and the engineering, chemistry and physics departments can come together around a table and discuss what we think prospective teachers ought to learn. That is something that has never happened before; we actually discuss curriculum. We actually do talk about it, and I think that is a step forward.

I would like to point out the effect on my colleagues. There is one other person in the chemistry department who is really interested and involved in this project. The younger faculty cannot touch it because they have to fight their way through the granting system for tenure. And almost all the people who are really successful in research tell me, without batting an eye, "You should not be wasting your time on this." They do not, however, interfere with the course at all. Why? Because, and I hate to state this publicly, but it is true — from the standpoint of most of the science departments, the elementary teacher education program is something no one cares about. But that gives you freedom.

Because if none of your colleagues cares, then you have the freedom to do what you think is right. I could not make such substantial gains in our own freshman chemistry program because there are too many vested interests. I can tell chemists, "Hey, if you want to do something interesting, then develop a course for elementary education majors, because none of your colleagues will care. And therefore, you can do something interesting and new."

I think we are being mildly successful. We now have three science courses required of the elementary education majors. Furthermore, the quality of these courses is improving because we are getting feedback from the students. The aura of fear has gone away. When we first started, students said, "Why do I have to take this thing? They did not have to take it last year." Some people fail, too. I failed eight people out of 80. You know, that was an unheard of thing. People failed! But now, it is a little different. Students walk in and they know there are standards in this course. They know that the laboratory is interesting. They know they have to go through it, and there is a different attitude. Pride is developing and it takes time to build that. I think my colleagues in the Education School are beginning to realize that I am not such an ogre as they thought I was, that I really care about those students and I am trying hard, in fact, putting forth my very best effort.

I honestly believe that all we talk about concerning curriculum frameworks and standards will come to naught unless the universities, which are turning out the teachers, begin to analyze themselves and begin to say, "Hey, we have to change the way we are training teachers." Because the teacher is the key to this entire revolution that everybody wants. I do not care how many scientists you bring in — I have visited middle schools and given liquid nitrogen shows and all that sort of stuff — but the minute I leave that classroom, those kids are back in the hands of the teacher, and we will never get away from that.

Now I would like to turn to the education of preservice secondary teachers. What are we going to do about those who are going to go out and teach in secondary school? They need different kinds of training, too. They do not need all the standard stuff that the ordinary chemistry major takes. In fact, I think many of those courses are even counterproductive for prospective high school chemistry teachers.

Many of our schools are rural. What does that mean? It means that someone who is educated in biology suddenly finds that she or he is teaching chemistry. I found out that many of the chemistry teachers in the state of Washington are not educated as chemists. In fact, often the best I can hope for is that they have degrees in biology, which means they have a modicum of chemistry, but certainly not the amount of chemistry needed to teach a laboratory-oriented science course safely, in an exciting way and with conceptual depth. They do not have that educational background.

Then you say, "Why don't we bring them back to the university? They already have degrees. Why don't we bring them back to the university and put them into a master's program to improve their chemistry?" The problem is we have a master of science in chemistry, but that is based on a four-year degree in chemistry. So, I look out there and I say, "Here we have all these teachers who are educated in biology, who are being forced to teach chemistry, and they have maybe two years of chemistry, possibly less. What am I going to do about that?" That is a serious problem.

What do these teachers generally do? They come back if they need a master's degree to raise their salaries. They come back, but what is available for them in the summer? Nothing, really. So they go to the Education School and take a few courses in education. They work on their degrees in educational administration or curriculum, which are the only opportunities open to them. Those degrees don't help them teach chemistry at all.

We have started another experiment. I wrote a proposal three years ago to the National Science Foundation asking for money to start a master of arts in chemistry, not a master of arts in teaching. The rationale was — we will bring practicing teachers who are out there educated in other subjects, primarily biology, to the university and put them through a rigorous program. And, in three years, using a month each summer, and home VCR for educational delivery, we will be able to offer a master of arts in chemistry, and educate them to teach chemistry the way it ought to be taught. I received about $700,000 from the NSF, and the Department of Energy agreed to hire the teachers to work for two months in the summer of the third year as chemists in the Battelle Pacific

Northwestern Laboratories. During that third summer, they will be taught biophysical chemistry in the morning by two-way television from the university and have the PNL scientists mentor them locally.

The capstone of this new degree program will be to write a thesis, essentially a master's paper, on what the teacher would do with all this knowledge and $5,000 to upgrade the curriculum, in his or her school or district. The program just started. And I am teaching the first course,

> **If you want really good staff development programs for teachers, design the programs for what they need and offer the courses through a university where a degree is involved, and make the incentives so high teachers will really want to do it.**

which is called General Chemistry from an Advanced Point of View. I have 25 teachers enrolled, and not one of them has a degree in chemistry. They mostly have degrees in biology, yet they are teaching chemistry in the state of Washington.

We have made the incentives in this program so high that many teachers will kill to get into it. I think this is the key. If you want really good staff development programs for teachers, design the programs for what they need and offer the courses through a university where a degree is involved, and make the incentives so high teachers will really want to do it. In our program they receive a $300 stipend per week while they are at the university. Room and board are also paid. All the credits are without charge. Furthermore, when they work at Battelle in a chemistry laboratory during the third summer, they will be paid as staff scientists for two months. It is going to cost about $29,000 per degree. You might think that is very expensive. Well, it costs about that much to incarcerate one criminal for a year in my state. If you think about it that way, it's cheap.

Over five years, we expect that 75-85 teachers from my state will eventually earn this master of arts in chemistry. They are already teaching chemistry, and through them we will substantially affect high school chemistry education in my state. Washington is not as big as California, but we represent only one university. Think of the impact if several universities sponsored such programs!

I have been describing the programs that one university has been sponsoring. Why aren't we doing more in my university and all the other universities? Because the university is a big part of the problem. I am not going to blame the K-12 schools for the teacher problems. We produce the teachers. The trouble lies at home. We do not have the right kind of attitude, interaction or developments in the modern university to solve the problems that we are helping to create in the public schools.

We are going to have to tell the scientists, who are very experimental and very daring in the laboratory, that they need to turn some of that effort toward being experimental and daring in the classroom and in the curriculum. Why are we so stodgy when it comes to that? We try all kinds of things in the laboratory. But when we walk into that classroom, we do the same old thing. One of my colleagues from Berkeley asked me, "Glenn, what do you want us to do? Do you want me to give up research? Is that what you want me to do?" And, I said, "No, no, no. I just want 15% of the time of the Berkeley chemistry faculty thinking about these problems. That's all I want."

I have spent a lot of my life in the last 20 years on these problems. I have not yet convinced most of my colleagues that what I am doing is right or worthwhile. However, I think time is on my side. For the first time, the National Academy of Sciences is really talking about education. We want to make sure that the National Academy of Sciences realizes that unless it considers the education of science teachers to be as important as the education of scientists, we will lose the game. I still have not convinced the Education School that I am not an ogre, but that group is coming around. We all have to start focusing our attention, not on our own little turf, but on what we are doing for the kids. And, when we start thinking that way, it is easier. It is a lot easier.

CITY SCIENCE — DEFINING A ROLE FOR SCIENTISTS IN ELEMENTARY SCIENCE EDUCATION

BY MARGARET CLARK

The City Science program, the elementary level of the SEP collaboration between the San Francisco Unified School District and the University of California at San Francisco, is in its second year of districtwide, hands-on, investigative science instruction in San Francisco's elementary schools. This partnership between a major research institution and a major public school district has provoked an examination of the role of scientists in the reform of precollege science education. We are still discovering how scientists can best contribute to this process and support teachers who are learning to teach science in a way that is exciting for them and their students. It's a little early in the game to provide a training manual for how to involve scientists, but it may be useful to describe what scientists have done in our program and some of the issues their participation has raised.

City Science began with a Summer Institute designed for 100 teachers, recruited from as many as possible of the 72 elementary schools in San Francisco. The first year, approximately 80 teachers attended the Institute, working in six grade-level groups with teams of one master teacher and two scientists. The teacher-scientist teams introduced each group of teachers to the kit-based curriculum unit chosen for their grade level. Their goal was to help the teachers become sufficiently comfortable with the materials and the science content to confidently take the unit back to their classrooms in the fall.

To prepare the scientists for their task, the City Science staff provided a three-day orientation just before the Institute began. During the orientation, the scientists received explicit preparation in the form of introductions to each other and their master teachers, a schedule indicating the times they would be needed, and the teachers' guides for the curriculum units. They also listened to speakers and discussed such topics as the stages of children's conceptual development and current reforms in teaching of mathematics and science. Less explicit, but perhaps more important, preparation came as they participated as students in sample activities. These included activities from some of the curriculum units as well as other hands-on experiences designed to illuminate the value of this approach to science instruction. During the orientation, the scientists were not told what they were supposed to do, except to work with their master teachers to take the City Science participants through the curriculum unit at their grade level. They were encouraged to define their own roles and to creatively supplement or extend the units to make them more effective.

Very quickly it became apparent that teachers and scientists work in different ways. The program was essentially set up from a teacher's perspective — the elements were outlined, but not rigidly defined. This approach allowed flexibility and accommodation to the needs of the participants as the Institute proceeded. Many of the scientists tended to be uncomfortable with the lack of explicit instruction or a well-defined role for them to play. Because most of the curriculum units were outside their specific areas of expertise, some wondered what they would be able to contribute.

This uncertainty began to dissipate at the end of the orientation, as the scientists began to work with their master teachers planning what they were going to do during the Institute itself. They were asked to write lesson plans and make a list of materials needed for each day, so the person in charge of materials had time to assemble them. They also went through the curriculum units, deciding which activities they would do with the teachers and what supplements and extensions they would add.

I worked as a scientist with the FOSS unit on Measurement for the 3rd grade level. Annabelle Shrieve, our master teacher, had gone through the unit in her own classroom. She brought with her a great many ideas for additional illustrative activities, and these gave my partner scientist, graduate student Anita Sil, and me ideas about what we could contribute. We suggested some extensions and went to the library to get information for their implementation. We also designed some adult-level measurement activities to challenge and stimulate the teachers, and we prepared written guides and materials for these. These activities included experiments that we conducted in our own laboratories illustrating how we use measurement in our specific research areas. Finally, Annabelle, Anita and I divided responsibility for leading the teachers through our selected activities and facilitating discussions with the teachers regarding what they were learning and how they could work with the unit in their classrooms.

Working with the teachers was a great deal of fun, and it gave me a very strong appreciation of their teaching skills and understanding of how to facilitate learning. I really looked forward to trying to implement what I learned from them in my own teaching of students at the university. The teachers, also not sure initially about what the scientists had to contribute, indicated pleasure at discovering that scientists could be friendly and could explain things in a straightforward manner. Tours of UCSF laboratories revealed windowless, crowded working conditions that dismayed the teachers but also provided a cultural view of the scientific community, with individual lab variations, that fascinated them. Essentially, they learned that scientists are people, and began to identify with them on that basis. A light-hearted camaraderie between the teachers and the scientists developed over the three weeks of their interaction.

The second Summer Institute modified somewhat the way in which scientists participated. Their involvement was for a period of only four or five days of the Institute, not including preparation time. Also only one scientist rather than two worked with each master teacher on the curriculum units. The number of curriculum units was expanded from six to eight, and teachers selected the module in which they wanted to participate.

In addition to the curriculum units, the Institute featured a workshop that gave all the teacher participants experience in asking questions that are appropriate for investigation and designing an approach to carry out the investigation. This workshop was based on the SPRITE (Science for primary teacher education) materials developed in the United Kingdom. Filling the first week of the Institute, the SPRITE workshop was conducted in four concurrent sessions by teams consisting of two master teachers and one scientist. At the end of the first week, teachers, working individually or in groups, generated their own questions which they then investigated during unscheduled periods in the remaining two weeks of the Institute. The scientists participating in the SPRITE module were available to provide advice and consultation. At the celebration on the Institute's last day, the teachers displayed posters on their investigations, showing a great deal of creativity.

The scientists who participated in the 1991 Summer Institute did not become heavily involved with the teachers during the subsequent school year. The program did not provide any explicit encouragement or discouragement for continued interaction. A few of the scientists did visit classrooms of teachers with whom they had worked in the Summer Institute, and they did a variety of things. Only a small part of their activities were directly related to the curriculum unit on which they had worked with the teachers. Primarily, they talked to the students about what they did in the laboratory and what it was like to be a scientist. In at least one instance, the teacher prepared the students the day before the visit, describing briefly what the scientist did and asking the students to prepare questions for him.

During the 1992 Institute, it was decided that teachers could benefit from a more formal arrangement in which scientists would be available to work with them in the classroom. Because the number of scientists who participated in the Summer Institutes is small relative to the number of participating teachers, additional university volunteers are being recruited to form teams that will "adopt a school" and engage in ongoing interactions to support the implementation of the City Science program. In the SEP tradition, the format of this interaction is not being stipulated. The City Science staff will help match schools and scientists and nurture these site-based partnerships as they develop a plan to support and expand the active approach to science teaching that City Science provides.

After two years of the City Science program, the scientists continue to define roles for themselves in elementary science education. The most obvious role is to be experts who provide evaluation and supplementation in science content and process skills while teachers are working with the curriculum units. Their broad knowledge base also allows them, as one of the scientists put it, to "demystify science" for the

---

A less explicit role, but one that teachers pick up on immediately, is that of showing that scientists are real people, often very sociable, with varied interests and a sense of humor. And their enthusiasm for science can be contagious, helping teachers enjoy science more.

---

teachers. One effective way they do this is by making connections between different areas of science or different scientific concepts. They also help teachers connect the science that is being studied with the real world and with the teachers' prior science education. Just as important, scientists help teachers learn how to appreciate the investigative approach, and how it differs from performing a prescribed series of laboratory exercises. In essence, they model all aspects of the scientific process for the teachers, generating questions, designing coherent pathways to find answers, and challenging the validity of results and the logic of the teachers' conclusions all along the way.

A less explicit role, but one that teachers pick up on immediately, is that of showing that scientists are real people, often very sociable, with varied interests and a sense of humor. And their enthusiasm for science can be contagious, helping teachers enjoy science more. Teachers also like seeing what a scientist does, so they can provide more realistic encouragement for their students who may be interested in becoming scientists. Even for students who don't want to be scientists, the university investigators can help teachers identify ways in which a strong science background can be useful in different areas of work.

These roles involve the scientist as teacher, but scientists also learn from the teachers. Most of the scientists who have worked with City Science have developed a high regard for the teaching skills, knowledge of children, and dedication of the teachers with whom they worked. The relationship that has developed between City Science teachers and scientists is one of reciprocal respect, with an understanding that the task they are mutually undertaking is well worth their time and dedication. This provides important support for the teachers as professionals. Further, when scientists go into the classroom, they become educated about the challenges and barriers that teachers face every day as they attempt to provide good science teaching. Once exposed, they can play an important role in the larger general and scientific communities in calling attention to these problems and becoming, as one teacher put it, "well-informed allies," whose professional standing gives them a broad influence.

Having participated in the City Science program, I have come to see that the creative tension between the scientists' and the teachers' perspectives can provide an atmosphere that facilitates discovery of what enhances the experience of learning science. It is essential that both sides enter into the partnership with open minds, willing to learn from each other and to work together as allies. The collaboration between these two groups, each bringing its special knowledge, strengths and vision, can then achieve what neither group could accomplish alone.

TEACHERS' COMMENTS ABOUT THE INSTITUTE

"Our unit leaders, scientist Christine Field and master teacher Christina Wilder, have done an excellent job presenting this material even though the topic was not in their field of expertise. They provided a direct model that a topic can be taught well even though the teacher may not be a scholar in that area. It's very exciting."

"It is great to have scientists around to bring authenticity and explanation and guidance. They also showed me what it means to scientifically ask and investigate a question."

"I enjoyed the laboratory tours very much. I found all the scientists to be happy we were there. They shared their knowledge without condescending, showing great ability to simplify abstruse information. I also appreciated their genuine interest in us. Congratulations to the scientists/presentors in the tours for helping to inform us and make us feel we are doing something important."

"Instead of tours, I would rather hear about scientists' early experiences that led them into their field of work — what teachers and activities got them involved."

"Great partnership — classroom tips from the master teacher, scientific information from the scientist."

"Access to the scientists and the university allowed for a valuable exchange of ideas regarding science and public education."

"What would I like to see added? A session where several scientists discuss/present their individual work, and how they communicate and cooperate with scientists in other cities, states, or countries."

SCIENCE AT THE CORE: SHEDDING NEW LIGHT ON PARTNERSHIPS

BY CARY SNEIDER, CHRISTOPHER DELATOUR AND KAREN MENDELOW

Two of the Bay Area's premiere science centers, the Lawrence Hall of Science and the Exploratorium, collaborated to provide three years of summers workshops for San Francisco's middle school science and mathematics teachers. The Lawrence Hall of Science is a public science center on the campus of the University of California at Berkeley. For over 20 years, Lawrence Hall staff have been developing and testing new activity-based science programs, such as Great Explorations in Math and Science (GEMS), that have reached millions of students nationwide. The San Francisco Exploratorium is an internationally famous museum of science, art and human perception. The Teacher Institute at the Exploratorium has been conducting intensive teacher training workshops in physical science for almost a decade. One product of these workshops, The Exploratorium Science Snackbook, provides complete instructions on how to build classroom versions of over 100 Exploratorium exhibits.

The lights have been turned out and the shades drawn at the Exploratorium Teacher Institute, a circular classroom insulated from the sounds of hundreds of excited children and adults investigating the Exploratorium's world-famous exhibits. Thirty teachers are clustered around six tables in the darkened classroom, faces alight with colors emanating from Luxagons, oddly shaped black boxes located in the center of each of the tables. The teachers are working in lab teams of two or three. Using filters, one team has isolated red, blue and green light, and is now adjusting small plastic mirrors to combine the variously colored lights to form new colors on a white cardboard screen. Another pair is discovering that rectangular and cylindrical plastic boxes filled with water can function as lenses. A third group of teachers is creating an artistic "light show" with small mirrors made of flexible mylar.

Imagining that we are in H.G. Wells' time machine, we whip through the remaining month of the first Science at the Core Summer Institute, and witness the teachers involved with dozens of hands-on science activities. Teachers create terrariums and use them to observe the relationship between plants and animals as well as the ever-important process of decomposition. They perform colorful chemistry experiments with purple cabbage juice, a natural acid-base indicator. Using devices that they have built at a cost of a few cents each, they measure the altitude of the huge pillars outside the Exploratorium. Teachers engage in a dynamic flurry of activity as they zip in and out with labeled boxes, carrying back to their middle school classrooms materials and resources for experiments that they have gathered and constructed in the workshop. Equally important are the many small discussion groups focusing on issues such as the integration of math and science, teaching strategies for students with limited English proficiency, or the logistics of doing more hands-on science instruction. The objective of these meetings is to devise action plans for integrating the workshop materials and experiences to meet the goal set by the San Francisco Unified School District (SFUSD): to make science an exciting core subject for all students at each of the district's 16 middle schools.

The Science at the Core project brings together the SFUSD, the Exploratorium and the Lawrence Hall of Science. Just as the three legs of a tripod provide equal and firm support, Science at the Core provides support by combining the efforts of a school district, a university curriculum development center, and a hands-on science museum. The important role of mathematics was recognized from the beginning. Several math teachers joined the program and several workshops focused on integrating mathematics and science.

With funding support from the California State Eisenhower Science and Mathematics program, we set out to achieve our goal by building strong teams of two to five teachers in each San Francisco middle school. Our plan was for the teams to be involved in the program for three consecutive summer institutes as well as frequent special workshops throughout the academic year. Having concluded its third and final major funding year, we are assessing the impact of the program, tallying up our successes and concerns, and planning for the future.

Although our evaluation is not yet complete, we can make a few generalizations. Strong science programs were indeed established at several of the middle schools. Some of the schools formed cohesive science departments that not only supported hands-on activities in the classroom but also conducted science fairs and other schoolwide science events, and coordinated the efforts of science and mathematics teachers. At other schools, individual teachers were successful in conducting hands-on science lessons with their own students, but were less able to enlist the enthusiasm of their colleagues.

We also encountered some unanticipated problems. Some of the most effective leaders left the school district to teach elsewhere in the country. While they are no doubt making important contributions to education, their impact is no longer felt in San Francisco's public schools. The biggest crisis was in the 1991-92 school year when financial difficulties forced the school district to send layoff notices to many of the teachers who had been with the program for three years. Fortunately, most of the teachers were able to keep their jobs, and morale has improved since then. However, the dire financial condition of our schools continues to pose great obstacles with minimal or nonexistent budgets for science materials, crowded classes, and lack of adequate facilities.

Some of the most important lessons learned during the program have come from the interaction between the staffs of the Exploratorium and the Lawrence Hall of Science. Given the various styles of teaching and learning, we found that it is important to respect and allow the other organization to try different approaches. A good example of this interaction is the light-and-color activity that introduced this article. When the activity was first presented by Chris, just one source of light in the center of the classroom was used. By holding prisms, mirrors and lenses in the beam, Chris was able to demonstrate spectacular light and color phenomena. With his background as a college physics professor, he was able to lead a fascinating discussion about the nature of light and color. Cary was frankly critical: "Demonstrations and discussions work well for teachers and motivated high school students, but middle school kids especially need to get their hands on things. They need to create the phenomena themselves, to play around. One light island in the middle of the room just won't cut it."

Chris took up the challenge. In a few weeks he had developed a simple but effective instrument that could be placed on every table in the classroom. Soon to be called Luxagons, the instruments have enclosures on four sides that allow pairs of students sitting around the table to work in teams. With the addition of more filters, mirrors, inexpensive prisms and lenses, there were enough materials for all the students to do their own experiments. As with all effective educational activities, the Luxagon has undergone many rounds of testing and improvement so that it is easy to use in the classroom. More than 1,000 of the devices have been distributed, and both the equipment and the activities continue to evolve.

Having finished its three-summer cycle of intensive workshops and major funding, additional sources of funds have become available for workshops and special science events so that the momentum and camaraderie developed during the project is not lost. Program teachers have taken responsibility for organizing continuing Science at the Core activities. Many teachers applied for their own grants to support activities in their classrooms and school. Several teachers have conducted workshops for their colleagues and have become resource mentor teachers in the district. Nonetheless, the teachers, school district, and staff development providers still face the enormous task of building upon this experience and resource base to make the kinds of systemic changes needed to radically improve the science education experience for all students.

SUMMER WORKSHOPS AND A MOBILE VAN AS CATALYTIC VEHICLES

INTERVIEW WITH LANE CONN

The Teacher Education in Biology (TEB) program, in its seventh year, is a joint project from the Departments of Biology and Secondary Education at San Francisco State University. Over 500 science educators throughout California have participated in TEB workshops, symposia and other activities. This educational program focuses on genetics, molecular biology, DNA technology (e.g., biotechnology) and its societal/ethical issues for high school and middle school science teachers, university teacher education faculty and district/county science coordinators. The TEB program has received funding from the National Science Foundation, the California Department of Education and California biotechnology companies.

The TEB program provides year-round program support and partnership development with participants implementing program topics in their classrooms. We begin with a two-week summer laboratory-based workshop and an applications- and issues-oriented three-day symposium on the science, applications and possible implications of biotechnology. Program activities then proceed into the academic year with follow-up renewal sessions, classroom visitations with materials and equipment, and other in-service opportunities for educators throughout California.

As an essential component to classroom implementation, the program provides the Helix I Mobile Outreach Laboratory, supplied with equipment, materials, and program teaching staff (for team teaching within participant classrooms) for visits to selected participant classrooms throughout California and for in-service workshop engagements. The program provides access to low-cost supplies and teaching resources through its Biotechnology Teaching Laboratory and Hotline, a newsletter for disseminating teaching-related information, and a variety of academic year activities for teachers and students.

Art Sussman: How did the Teacher Education in Biology program get started?

Lane Conn: In 1986, David Micklos and the Vector Van from Cold Spring Harbor Laboratory gave a one-week workshop on DNA Science for California high school teachers at the University of California at Davis. Thirty-six teachers participated. At the conclusion of the workshop, the teachers were fired up to bring DNA science into their classrooms. Unfortunately, they were all alone, without any California infrastructure and support for their implementation dreams.

The California Department of Education asked me to observe the UC Davis workshop and gauge its effectiveness for biology education. The Cold Spring Harbor workshop at UC Davis demonstrated a high level of educator interest in the field of DNA Science and launched me on the road to developing the TEB program. After working in the biotechnology industry for five years, observing the educational needs at the UC Davis workshop and talking with teachers, it was obvious that a great need existed in the public (i.e., citizen) understanding of biotechnology's realities and potentials. The schools seemed the place to start. With further investigation, I found that no mechanism to disseminate accurate and relevant information was available in public education. We initiated the TEB program to meet this documented need.

Professor Crellin Pauling, then chairman of SFSU's Biology Department, Dr. Kathleen O'Sullivan in the Department of Secondary Education, several other scientists and educators, and I developed a proposal for the National Science Foundation to do staff development with teachers concerning the scientific and societal/ethical aspects of biotechnology.

AS: What did you plan to do with the teachers?

LC: The first idea was simply to offer a two-week workshop. I had seen one-week workshops in action and thought they were much too short to address the critical issues. As we saw it, the critical issues were to give high school teachers a knowledge base and laboratory experiences that would focus on molecular biology and molecular genetics. It would link their lessons on classical genetics to the more modern aspects of DNA and molecular genetics. We discovered that although almost all the teachers cover classical Mendelian genetics in their curriculum, they came to the workshop with less understanding than we had expected about the larger perspective of Mendelian genetics and how it related to human genetics, molecular genetics and DNA. So we realized the necessity to start our workshops with fewer assumptions about teacher knowledge and experience, and spend some time with the fundamentals. We also had to begin the societal/ethical interdisciplinary instruction from scratch. Nobody seemed to be doing these things together.

AS: So there is a content section?

LC: Yes, the workshop has a content section that begins with the fundamentals of molecular biology and grows into molecular genetics and recombinant DNA. A one-week workshop is simply not sufficient time to practice the experiments and address ways of implementing this material in classrooms. The one-week classes that I had seen at UC Davis did not do that at all. They were just very intent on learning molecular biology, and did not seem as interested in addressing classroom implementation. There is a huge amount of information about how to implement, how to structure and sequence the lessons, and what your expectations could and should be for students.

AS: Even a workshop that focused on doing hands-on experiments with teachers would still benefit from an implementation section. You don't just practice doing experiments; you also have to figure out the classroom strategies.

LC: That's exactly right! We have learned that there are many little hurdles all along the way that impede teachers from bringing new, relevant and exciting curricular issues into the classroom. This is not to blame the teacher because many of the hurdles are out of their hands. When you directly confront those implementation hurdles and lower the barriers that prevent people from really bringing new experiences into their classroom, then interesting things start to happen. We are beginning to get information about what works and what does not work. One feature that facilitates implementation has been the development of partnerships between participating teachers and scientists from local industries, universities and community colleges. In most cases, these partnerships are led by teachers and aim to bring hands-on experiences in modern biological sciences to students.

AS: Many programs throughout the nation seem to be dealing with this issue. Teachers will go to summer workshops and have an experience that is undoubtedly beneficial to them. Anybody who talks to the teachers or just observes what is happening realizes that these experiences enhance the professional skills and status of teachers. Yet, in terms of what they bring into the classroom, in general the results seem to be less than we would hope.

LC: Yes, but it is not that simple. This is our seventh year of conducting the TEB program, so we have six years' worth of teachers out there throughout California. They cover the whole spectrum from "I haven't implemented anything" to "I developed a multi-partner relationship with other teachers and scientists, we have taught the lessons in many classrooms, and we have developed new experiments." We are beginning to analyze what aspects of support, partnership and what aspects of school-based or district-based realities have assisted teachers or prevent them from implementing the lessons effectively with their students.

AS: Which brings us to the segment of TEB with which you are identified — the Helix I van.

LC: Students often call me the van man. The Helix I Mobile Laboratory was part of the program concept from the very beginning. I wanted the program to go beyond simply a summer workshop experience and make sure that the material directly impacted students. We needed to get equipment and supplies directly into the teachers' hands at their sites and I needed to see for myself the realities of their teaching environments. I also believed that it was important for program staff to have direct experiences of team teaching and observing these lessons in high school classrooms. The Helix I van annually serves a selected number of participants, its visit does not cost the participant or the school, and it makes only one visit to a teacher's class. After that, we work together to make these resources available locally through the development of partnerships.

AS: We need to talk about the Helix I van.

LC: The van has a set of equipment and supplies in it that allows teachers to implement experiments in molecular biology and molecular genetics. These are sophisticated materials that are not normally found in high schools. I drive the van to school sites, deliver the equipment and supplies, and help the teachers set up coordinated experiments and lessons in biotechnology.

The van comes as a total package, and I think that is a fairly critical aspect. The teachers are the instructors and they have to be concerned about teaching. When the van comes with its set-up of 25 boxes of equipment, many people get confused and think that it is like a bookmobile that students visit. It is not that. The van transforms the classroom into a laboratory where students can perform modern scientific experiments. It brings everything a teacher needs: petri dishes, tissue cultures, bacterial plates, micropipets, enzymes and sterilized media for growing bacteria. It has all the experimental supplies for classes containing up to 20 teams of two students. So the teacher does not have to worry about "where am I going to get X, Y &Z." The teacher can focus on doing the instruction. Remember, these are all teachers who have gone through the summer workshop.

After the van shows up, the program will stay at the school for approximately four weeks. In the first year of the program, we started with one-week commitments and I drove the van to about 32 sites across California in one-week stints, and that was really fun. However, we realized that one week is not enough time for anyone to effectively implement the labs that many teachers and students want to do. In that first year no one attempted to make a recombinant and do the whole set of experiments that they had been trained to do during the summer. There was not enough time. As the program matured, it went from one week in the academic year to a deeper and longer classroom experience. Now I think it has matured to the point where the van serves a smaller number of schools with a four- to five-week commitment per school.

AS: Are you at that school site during that whole time?

LC: No, I, or Ann Moriarity, the new TEB outreach coordinator, go with the van, unload it, and help the teacher set things up in their school. When I get there, it is a partnership between TEB/SFSU and that school. The teacher does not go off and take a coffee break while I set up. We work together to get it all laid out in order for them to understand exactly what they are getting. An inventory system comes with the kit, so it is very clear and they can find everything. I stay there usually a day to begin the team teaching with the teacher. Then I leave and it becomes the teacher's responsibility to conduct the instruction. We have found that the most successful schools are those that have a team of teachers who are working together — two, three or even four teachers that can share the responsibilities and the goals.

AS: So the van stays there?

LC: I drive the van home, but all the materials stay in the classroom. I unload all the materials and the van goes away usually that same day or the next.

AS: So if you had more copies of materials . . .

LC: First, I want to say that Genentech Inc. contributed the van and all the equipment and materials in 1987, before we had secured any additional funding. Not only did this contribution give us the physical materials that we needed, it also helped us secure funding from the National Science Foundation to run the workshops and the program. I think it was a real selling point with NSF and the California Department of Education because they saw a private sector company that was willing to make a substantial contribution.

Mobilizing only one kit under-utilizes the van. I was given a second kit by a Sacramento science education company called Grau-Hall Inc. One person mobilizing two kits and running a program is effective but exhausting. I learned that although one van could effectively mobilize two kits, that alone would be a full-time job. It takes a van outreach coordinator to have the sole responsibility of just keeping the van going during the academic year.

AS: What is needed now in order to impact more California students?

LC: What the students and teachers of California need now are more partnerships between schools and local businesses, industries, colleges and universities with the objective of supporting quality classroom science instruction. Whether it be biotechnology or conservation biology or organic chemistry education, the specific content is not as important as the involvement. All these organizations can contribute human and financial resources to help improve the future for our students, which will improve our future as well.

AS: What do you think is the most significant lesson that you have learned from using Helix to service teachers in the schools?

LC: The most important thing that we discovered about the van is that it becomes a catalyst that excites and convinces local groups to actually implement new teaching practices, and then develop local support structures (e.g., partnerships) so the instruction continues within their

---

The van becomes a catalyst that excites and convinces local groups to actually implement new teaching practices, and then develop local support structures

---

local environment. We have learned that there is an energy barrier to implementation. Teachers may worry that their students will not acquire the skills to perform the techniques that are necessary to do the experiments. They may have fears that the concepts are too advanced. Once the van has visited the school and they have implemented this new and exciting instruction, most of their fears leave and teachers are then motivated to find other, more permanent, resources to keep the process growing locally, and that means partnership with businesses, industries and local colleges and universities.

The van significantly lowers the energy barriers of not having the materials and not feeling confident about the teacher's or the students' abilities. You cannot just have a van dump off a bunch of equipment and go home. You need a driver who has teaching experience and is committed to actual implementation. Once the catalysis has happened and teachers feel confident that they can teach and their students can learn these aspects of biology, then the teachers decide that this has to happen every year. Since the van can visit only a limited number of schools for a limited period of time, it catalyzes the teachers as well as members of the administration — and hopefully local colleges, universities and

---

industries — to work together to make sure that this kind of instructional package remains continuously available for students.

AS: And that is what happened in San Mateo County?

LC: Yes. That is what happened in not only San Mateo County, but here in San Francisco and is developing in at least four other counties. Ms. Suzanne Black, Mr. Stan Ogren and Ms. Kathy Liu, all San Mateo County high school teachers, Gary Nakagiri (county science and math coordinator) and I met with the Genentech Foundation to seek funding to purchase a kit of equipment and materials that would remain in San Mateo County and circulate among the high schools in the county. Dr. Herb Boyer, one of the founders of Genentech Inc., and other prominent scientists were present on the foundation board. This meeting was a great experience. The board members did not want to hear much from Gary Nakagiri or myself. They just wanted to hear the teachers tell them what they needed in order to do their job well and teach their students. The three teachers gave short presentations, answered questions and the foundation board granted the money. It was a thrilling meeting. The teachers experienced that scientists were committed to their cause and wanted to support them in their efforts.

AS: In other words, we are talking about making these materials available continuously on a county level rather than relying on the Helix van's equipment.

LC: Yes, it is making a transition from a temporary resource to a permanent partnership resource that the teachers can control. The teachers do not have control over TEB's Helix I van. But the partnerships which Helix has catalyzed can easily adapt to local environments and local needs. The teachers can write grants and work in partnerships in the local area. These partnerships can involve higher education institutions and industries to work with teachers and take this partnership program in directions that Helix has never gone. As an example, Marin County teachers and scientists have developed experiments and a kit-based partnership that is totally different from what we use in Helix .

AS: Another nice feature of the van is that it can go and service teachers at locations that are rural and hard to reach.

LC: The van has gone to some very rural settings in Northern

California, in the Sierra Nevada and in the Central Valley. The rural teacher has a much more difficult challenge of implementation than a teacher in the S.F. Bay Area. My impression is that these rural areas do not have the district infrastructure or sufficient networking and contacts within their region. Although rural kids can learn just as effectively and are just as motivated by things they read in the newspaper and see on television, the resource network for teachers is not as readily available. The Helix I van has visited these rural sites and we have helped develop local connections between industries, the U.S. Forest Service, or local community colleges.

AS: One other thing that has come to my attention with respect to biotechnology is integrating the societal issues into the program. Does the van help with that?

LC: The van helps to implement the laboratory aspects of our program. Only in a small way has it assisted in implementing examination of the ethical and social issues that we instruct in the program. This is an area we are currently working. I have observed that it is easier for biology teachers to do the "straight science," than the related social issues. If you understand the content and you have the equipment, you can do experiments that can get off the ground fairly easily. The students can understand the lessons and it is a focused program. When you then try to address some social ramifications of the technology, that brings in social science and an open-ended analysis of the affects of science on society. Although it is completely anecdotal, my observation is that science teachers, which I am one, are more comfortable doing DNA electrophoresis than discussing its social implications.

AS: It seems to me you run into the same barriers but they come in a different form. In order to do an experiment you need to have the equipment and supplies. In order to do a social analysis, you also have to provide supplies — but this time they are information packets, essays, laserdiscs and videotapes rather than electrophoresis gels and micropipets.

LC: I agree that one of the challenges is that teachers do not have access to these materials. They also do not have effective teaching models. Turning on a video and telling students to take notes is not an effective technique. Teachers need techniques to foster interactions with the electronic media. They also need the context for analyzing the social issues and that is one area that we cover in our symposium. At our symposium we explore one or two selected applications of modern molecular biology and use the best models out there for ethical issues instruction — short vignettes, case studies, simulations and the Ball State Model for Bioethics Decision Making. The short vignettes are something you can do in a few minutes in the classroom. The simulations may take up to a week if the teachers have the information resource materials already available.

AS: We had not discussed the symposium. What is it?

LC: In a typical summer we have four workshops with about 25 teachers at each one. The symposium collects all of those participants together at one site at the end of the summer. The symposium is a three-day event as opposed to the two-week laboratory-based workshop. The symposium is where we address the social and ethical implications of biotechnology.

AS: Participating teachers commit to a workshop for two weeks and then a symposium for three days at the end of the summer. Do you have problems recruiting enough teachers?

LC: We do have an issue of recruiting teachers because there has been a drop in enrollment. Some people believe we are tapping out a certain region in terms of its teacher pool. I completely disagree with that. I think the majority of teachers have never taken our workshop or any workshop in biotechnology, but how do you serve them? How do you understand what they will take advantage of and how a program can serve their needs. One issue, for example, that we have run across in the last two years, which continues to be a frustration for me, is that women call me and say, "I really want to take this program, but I have two kids and I would need to pay for day care." I have tried to talk to NSF about this. Programs like ours that are multiweek programs over the summer should provide some support for parents, especially single parents, of young children.

AS: But the teachers are getting stipends.

LC: The teachers are getting stipends of $60 per day, but that is a minimal summer income. They do not feel if they take that stipend and turn it over to day care, that they are really getting compensated. This is just one issue that prevents teachers from taking advantage of summer programs.

AS: Some, perhaps many, teachers do not want to do in-service during the summer. They may have another job or want to spend time with family.

LC: That is why programs have to offer opportunities during the academic year, possibly through districts or counties using in-service programs to begin that catalytic process of staff development to get them engaged. I do not mean two hours a month. One model is to provide substitutes and have the two-week in-service during the school year. We are working toward that now.

AS: You said that some of the teachers who take the summer workshop come back and team teach the following summer. Who is the faculty during the summer workshop?

LC: We have been encouraged to increase the amount of teacher leadership development in our program, through teachers teaching teachers as one model. During the summer there is a five staff faculty at each workshop; there are two lead teachers, two university faculty members, and a laboratory technician. The two lead teachers have participated in the program previously and taught it to their students, usually with Helix support. The teachers are responsible for conducting and administering all the labs. The program is about 70% lab-based and the two lead teachers are responsible for all the laboratory instruction. The scientists will be in the lab along with the lab assistant, but the lead teachers are the managers and organizers of the laboratory.

AS: How do the scientists help during the summer?

LC: Scientists help with the content base of the workshop. They give morning lectures and help with afternoon discussions, but the real bulk of the effort is made by the lead teachers. One important thing is that a number of scientists who have participated at the different workshop sites have told me that they are better university instructors now that they have been able to watch exemplary science teachers during those laboratory sessions. I know that the instruction of the undergraduates at SFSU has been enhanced by those faculty who have participated in our

TEB workshops.

I have been very surprised that faculty members in the program rarely go into the precollege classroom. Over 90% of them have never set foot in a high school classroom even though they are teaching within the program. This year we have a new mandate — the university scientists from the summer workshop have to spend one day a semester going out to a local teacher's classroom. I don't care what they do — they can just go and watch. I want them to experience the challenges in public education so they will understand it better. I think that can help catalyze a greater interest in working with our schools and teachers.

AS: Another catalytic effect is that you sponsor the workshops at different campuses.

LC: That is a major goal of this program, to help other California universities develop local staff development programs. San Francisco State University cannot serve all of California in biotechnology education, but it may catalyze Fresno, San Diego, Los Angeles, Santa Barbara, Santa Cruz, and Davis into offering more programs that work in staff development during the summer and academic year. We also want to catalyze effective outreach programs directly into the classroom from local colleges and universities.

AS: Have you seen that happen?

LC: We have seen it in some degree in Sacramento, Fresno and San Diego after the workshops happened there. I think the biggest limiting factor is simply sitting down, putting a proposal together and getting it funded. The barrier is not motivation of the faculty or the teachers as there has been a network developed at all these sites. If the sites were funded, there would be no question of their success.

AS: Any final comments?

LC: I have seen the equipment being used in the classrooms of teachers that have not taken any staff development in this area. Sometimes, the van is at X school and three teachers who participated in the workshop are using it, and there are other teachers over here who see the van and ask, "Can we just do a little electrophoresis with our students?" And the experienced teachers are often willing to work with them. That is an interesting form of catalysis. I think the percentage would be very high for those novice teachers to then take a summer in-service workshop. So, I would love to see a model of the van that would contact science teachers who have not taken the workshop. We would work with them for a day or two and then recruit them for the following summer. This mechanism could help us address a common problem — how do we get teachers who do not normally attend extended in-service programs or who resist learning new classroom practices? Many of us are concerned that we need to see new faces at the workshops. The active teachers who jump at every opportunity are great, but we need to reach the other teachers as well.

I also have to tell you about one teacher who has shown me a whole new aspect of Helix. He was greatly disappointed when he took the workshop two years ago and Helix was not able to visit him in Martinez. His name is Lou Pruitt at Alhambra High School. Helix was visiting his lab partner from the summer workshop so he took four days off, got a substitute, and drove 60 miles away to Joe Lynch's classroom in Lincoln and team taught his classes with him. These were not his kids or his school, but he was determined to experience the program with real students. Then he went back to his school, wrote about 20 grants and received more than $17,000 so he could implement the program in his school.

He helped me see another role for the van. Helix is not a resource for all teachers and it cannot serve all the teachers that go to the program, unfortunately. But when it goes to a school, it can look at a 25-mile radius around that school and see what other participants could benefit from the experience of coming and maybe team teaching or even just observing the implementation of the program. Although Helix cannot serve every school, this is another way it can act as a catalyst.

BAY AREA SCIENCE & TECHNOLOGY EDUCATION COLLABORATION

ADAPTED FROM AN ARTICLE WRITTEN BY KEN EPSTEIN

As a veteran high school science teacher, Beth Napier had come to expect that she was on her own when she needed new curriculum for her classroom. She assumed her financially strapped urban school district would be unable to provide either resources or in-service training to help her develop new physics and chemistry lessons.

Recently, however, opportunities have begun to open up for Napier, who has discovered that she can tap into state-of-the-art science and technology. "I spent a summer doing some research for the next Mars flight. My whole image of what scientific research is and what scientists do has changed," Napier said. She now also has access to expensive equipment to use in her high school lab and receives hands-on experience using the latest curriculum. "One of the workshops I took was called Global Warming and the Greenhouse Effect, where I learned up-to-date information and was given things I could use in labs at my school.

"I'm being exposed to new ideas; I'm trying new things; and I'm getting results in my classroom."

Experiences similar to Napier's are being replicated in elementary and secondary science classrooms throughout the Oakland Unified School District, the sixth-largest district in California. It is the result of a unique, new collaborative relationship between the district and four of the nation's most prestigious national research laboratories. Called the Bay Area Science & Technology Education Collaboration (BASTEC), the partnership is coordinated by Lawrence Berkeley Laboratory.

The other three national labs are Lawrence Livermore National Laboratory; Sandia National Laboratory, Livermore; and the Stanford Linear Accelerator Center — all in Northern California. The partnership also includes the Lawrence Hall of Science, Holy Names College in Oakland, California State University, Hayward, and other educational institutions, as well as organizations such as the Edna McConnell Clark Foundation, the African Scientific Institute, and the National Organization of Black Chemists and Chemical Engineers.

At a time of declining school funding, BASTEC provides the school system with at least one pillar of financial stability. When the project began in the fall of 1990, the U.S. Department of Energy backed it with a $500,000-a-year grant for at least five years. Rather than working directly with students, BASTEC has focused its energies primarily on in-service training and providing teachers with resources, based on the premise that each teacher who is aided will eventually reach hundreds of students.

The project supplies a science center where teachers can check out curriculum materials and expensive equipment. It offers mini-grants up to $1,000 for teachers to develop and disseminate curriculum, organize field trips and perform science projects in their classrooms. It also provides summer workshops taught by teachers and scientists, for which the Oakland teachers receive a stipend.

BASTEC recently held its second annual one-day mini-conference for the Oakland school district. More than 450 teachers attended the 40 sessions and workshops, which covered topics as diverse as "The Tropical Rainforest," "Popcorn! Eating Is Science," and "Ecology Games for the Classroom."

In the short time it has existed, Oakland teachers and administrators say BASTEC has made a deep impression on the Oakland school district, working with nearly 700 teachers at more than two-thirds of the district's 90 schools and generating an excitement that is all too rare nowadays among overburdened and harried teachers.

"We're upgrading the teaching of science in our district," said Robert Newell, Oakland's assistant superintendent for Curriculum and Instruction. "BASTEC is an excellent resource for the school district. It's really an example of what a collaborative effort should be."

Because of its initial successes, BASTEC is already planning to broaden its goals and has begun taking on new responsibilities within the Oakland school district. The project has begun helping the district develop a new kindergarten through 12th-grade science curriculum. When completed next school year, the curriculum will explain which concepts should be taught at each grade level and let teachers know where they can find information and resources to teach the concepts.

BASTEC also has plans to create a "student pipeline" to help interested students enter scientific careers. The program will provide counseling and other types of academic support so students will be qualified to study science at colleges and universities.

A key reason for BASTEC's strong start has been its willingness to take the time necessary to build a consensus before initiating its first project, according to Roland Otto, Director of the Center for Science and Engineering Education at Lawrence Berkeley Lab.

"It took almost a year to get our mission and goals fine-tuned," he said. "Taking that much time was frustrating because we were anxious to start. But it meant that a lot of discussion could be set aside once the money arrived and we were ready for action."

Manuel Perry, one of the founders of the project, who serves as director of Education Programs at Lawrence Livermore National Laboratory, underscored the need for careful planning. From the very beginning, the national laboratories wanted Oakland to be the first district with which they would work, Perry said. The Oakland school district has 55,000 students and 3,000 teachers. The student body is 57% black, 9% white, 18% Asian, and 24% are classified as limited-English speaking. Students come from homes where more than 50 different languages are spoken, including Cantonese, Arabic, Tongan and Tigre, a language of northern Ethiopia.

While emphasizing careful planning, BASTEC's approach has also stressed flexibility and the need for long-term involvement to make lasting changes within an educational system. "It's not a get in and fix it situation. There are no magic bullets," Otto said. "We've never been in partnership before with an entire school district. It is important that we find a role for ourselves that will let us get involved and stay involved."

Otto also emphasizes the importance of other factors. While $500,000 per year for five years is not an enormous amount of money for a school district the size of Oakland Unified, it is enough to motivate key players to develop and implement collaborative plans. In addition to its grass-roots base, the program has always benefited from the support of contacts high in the district's administrative structure. The coordinating group made a crucial decision in making the program available for all schools and all grade levels. Otto believes that there is a strong temptation to focus efforts at particular schools or certain grade levels. While this approach may yield quicker results in "demonstration" sites, it raises many political issues regarding which schools and grade levels are selected. BASTEC's inclusive format avoided divisive political struggles and also is more in keeping with the goal of promoting systemic change throughout the Oakland Unified School District.

BASTEC funding pays for a program coordinator who is hired by the school district in consultation with Lawrence Berkeley Laboratory. In an interesting twist, this school district employee works in an office at the laboratory. Otto believes this increases the coordinator's effectiveness and stature. Other employees hired with BASTEC funds include three school district half-time positions for teachers on special assignment in science, mathematics and technology.

Oakland Board of Education member Darlene Lawson said district staff have been impressed with the project's approach. "It's more of a cooperative relationship with the schools than in some partnerships," she said. "They find out what the schools need rather than just coming and saying this is how were going to do it."

In fact, BASTEC has relied on the views and experiences of teachers at all levels of the planning process, said Eileen Engel, Pre-college Program coordinator at LBL. "What BASTEC does is get directly in touch with teachers, and ask: 'What do you want?' Teachers themselves decided what workshops they need and how they want to spend the grants," she said.

"They're in the trenches every day. They know what they need." BASTEC's initial activity, holding a meeting where 120 teachers brainstormed about their classroom needs, was met with a level of skepticism that is common among teachers across the country. "Their attitude was, 'We're here; we're going to tell you what we need: but we don't believe you'll do anything," Engel said.

The project's concern and sensitivity to teachers' views has paid off in teacher energy and enthusiasm. "The plan is not something written in concrete," said Napier, the high school science teacher. "Teachers are encouraged to take charge. I don't believe there are too many programs like this any place in the country."

Even with its openness and careful planning, BASTEC would not have existed without the Department of Energy grant, which enabled the project to translate teacher plans into practical reality. "It's our belief that if you make the resources available, teachers will take advantage of them. Some teachers are better than others, but through BASTEC we try to provide resources to every teacher seeking to improve," Otto said.

"Teachers were impressed when they planned something, and found that there was money and organization to guarantee that it happened the way they planned it," he added. "They're not used to that. We begin to give teachers a sense of hope, that they are not going to be alone."

One advantage of the way the resources of BASTEC are managed is that they remain independent of state and local funding variations, which often cause so much disruption to school district plans. BASTEC is able to keep the focus on science and technology education through good and bad financial times.

Besides money, each of the partners has brought unique strengths and resources to the collaboration. For example, at the Berkeley lab, Otto said, "One of our institutional strengths is that we know how to make concepts and ideas a reality. We can get supplies to teachers in a matter of weeks because that's what our researchers expect in their labs."

The national laboratories also have donated some basic materials after discovering that teachers had been working almost without supplies,

---

Engel said. "We found out that they needed colored chalk and graph paper. One teacher organized a fund-raiser to buy frogs to dissect. Teachers were lucky if they had $200 a year to run a biology lab."

Another partner, Cal State University, Hayward, has organized "Focus: Physics," a four-week residential summer program for 40 Oakland middle school students. The students participate in field trips to all four national labs and receive intensive precollege advising.

Though all of BASTEC's programs have proven valuable to the district, its work with with elementary teachers may be the most important, according to Helen Quinn, theoretical physicist and science education officer at the Stanford Linear Accelerator Center.

"Elementary teachers are expected to teach every subject," she said. "But some of them not only don't know science but are actively afraid of it. They can't challenge student to think about science because they were badly taught themselves. "To break that cycle, we show them methods and materials they can use in their classroom. They rave about the workshops."

Hawthorne Year-Round Elementary School is one Oakland school that took advantage of the project's resources to inject new life into its science curriculum. With a $900 grant from BASTEC, it purchased 475 sanitized owl pellets — little balls of fur and bones regurgitated by an owl after it has devoured its prey. To the uninitiated these pellets many seem a little repugnant, but to children they are treasure chests containing tiny skeletons of rodents, shrews, moles, birds and crabs.

The school utilized the help of a professional science consultant, who taught the hands-on lesson to 15 classes (ranging from kindergarten to sixth grade), while at the same time modeling innovative methodology for the teachers. Using cross-age and cross-cultural approaches, younger children were teamed up with older "buddies," and many of the school's limited-English speaking students were paired with others who were more proficient in English, who served as translators.

Sarah Shaffer, an independent science consultant to the school district who is gaining fame as a local "owl pellet expert," said students were enthralled by the lesson. "It's a no-fail lesson. Many students asked to skip recess and lunch so they could keep working. But the most significant effects were those they had on the teachers," she said. "Teachers held a brainstorming session to figure out how to integrate the lesson with the rest of their curriculum. In addition, at the same time I was teaching the students, I was modeling the lesson so teachers could do it again in the future. It's one thing to see a lesson written up. It's another to see it come alive."

The grant and the experiment it funded has had a lasting effect on the school and its staff, according to teacher Gail Whang. Teachers are using innovative methods and planning schoolwide projects. Currently, all classes in the school are studying oceans.

"The BASTEC experiment was a springboard for us," Whang said. "Now, we're trying new things all the time. It's amazing how much mileage you can get out of some owl pellets."

PARTNERS IN IMPLEMENTING CURRICULUM CHANGE: THE PITTSBURG MODEL

BY JACQUELINE BARBER, CARY SNEIDER,
KATHARINE BARRETT AND STEFAN GAIR

Implementing curricular change is a tremendous challenge. Current trends in science education point to the need for science programs that rely largely on activities, enabling students to develop science thinking skills and an appreciation for the nature of science. Further, the call is to help students construct a larger view of science through helping them understand some of the big, recurring ideas throughout all the disciplines of science — what's referred to as a "thematic" approach to teaching science. Constructing a curriculum that does all of this and yet builds from grade to grade is unique and half the challenge. The other half is devising a process for creating that curriculum that involves teachers, because they will be the users of that curriculum. Following is an account of a model for implementing curriculum change that is the result of the collaboration among school district personnel, teachers and university curriculum specialists. We call it "The Pittsburg Model."

Like most school districts, Pittsburg, Calif., was in need of an articulated science curriculum, reflecting the new state and national guidelines for science education. Science department staff in the two middle schools had worked together to create a well-coordinated, activity-rich, thematically based science program. However, effective science education in the elementary schools was spotty. Some excellent teachers had crafted good science programs in their individual classrooms, but for the most part, teachers relied on textbooks and presented very few hands-on activities.

Hit-or-miss inservice workshops wouldn't do. Stefan Gair, Science and Mathematics Curriculum coordinator of the Pittsburg Unified School

> # THE GOAL
>
> To design and implement a K-5 core science curriculum for all elementary schools in Pittsburg, California
>
> • taught by all elementary teachers
> • articulated by grade level
> • prepare students for middle school science

District, wanted to reform science education in a way that would result in long-term change. He contacted the Lawrence Hall of Science on the University of California at Berkeley campus for assistance. LHS is both a science museum for school groups and the general public, and a center for the development of science and mathematics curricula and teacher education. Given the many educational programs developed by the Hall, and Steve's experience in involving teachers to bring about educational reform, we seemed to have real potential for creating a successful model for change.

The overall goal was clear: to design and implement a K-5 core science curriculum that would be taught by all elementary teachers and articulated in such a way that students would not repeat units. It would prepare students for the thematic, hands-on program that they would encounter in middle school.

If we were to achieve our goal, we knew that it was important for teachers to participate in the decision-making. One reason is to ensure that the new curriculum plan speaks to the needs of teachers. Secondly, we wanted the teachers to be co-owners and creators of the plan, rather than feeling that it was being imposed. After all, they would be the end users! Although we would not be able to include all of the teachers in the district in the planning process, we would be able to include individual teachers who are enthusiastic, knowledgeable, and have the respect of their colleagues.

Our past experiences had made us aware of the tremendous power of site-based teams. We knew that for a science curriculum to be implemented and maintained, each site must have at least one teacher at each grade level who is skilled and confident in presenting the core curriculum, and who is willing and able to be a leader in science at the school. Building that skill and confidence would require intensive inservice workshops for each science unit in the curriculum. It is also important that all of the materials needed to teach each science unit be located at each school, and for one person to be in charge of maintaining the kits.

And finally, we had to be prepared to answer teachers' questions about how a hands-on science program would prepare students for examinations in science knowledge. Fortunately, California is one of approximately 10 states that are implementing an activity-based assessment program to determine whether students have acquired science process skills, such as the ability to observe and record data, to design experiments, and to draw valid conclusions. To link the new curriculum we were planning with new statewide assessment methods, we also had to develop effective assessment instruments for each of the units to be included in the curriculum. The instruments would enable teachers to determine how well their students were improving in their science skills, and in their grasp of fundamental science themes.

Pittsburg has seven elementary schools that feed into two middle schools. To achieve implementation across the district, we wanted to involve teachers from each of the elementary schools. The Science Task Force was formed, including a lead teacher from each school for each grade; a mentor from one of the middle schools who helped to create the middle school science program; three curriculum specialists (life, physical and earth sciences) from the Lawrence Hall of Science; and the project leader, Stefan Gair. The Task Force's

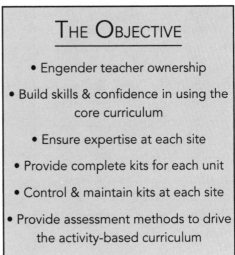

THE OBJECTIVE

- Engender teacher ownership
- Build skills & confidence in using the core curriculum
- Ensure expertise at each site
- Provide complete kits for each unit
- Control & maintain kits at each site
- Provide assessment methods to drive the activity-based curriculum

charge was to design and implement the curriculum.

We decided on a curriculum plan that included four science units at each grade level, each approximately three weeks in duration. While 12 weeks of instruction may not seem like a lot, it was 12 more weeks of hands-on science than many teachers were doing. It would, therefore, be a stretch for some. On the other hand, teachers who already presented a collection of their own exemplary units would have time in their curriculum to continue presenting their favorite units.

The Science Task Force aimed at a balance of units in the earth, life and physical sciences. The units would develop students' science skills and an understanding of the fundamental themes that underlie all of the sciences. The curriculum plan would include units that built students' skills and understanding during a single year, as well as from one year to the next. By the time the students graduate from elementary school, they should be sufficiently skillful and knowledgeable to be successful in middle school science.

THE PRODUCT

- 4 units per grade per year
- Each unit approximately 3 weeks in duration
- Balance of earth, life, physical sciences
- Skills and themes build within each year
- Prepares students for middle school
- Each unit includes assessment module

THE PROCESS

- Establish Science Task Force
- Survey all K-5 teachers about science teaching. What? How? How much?
- Task force teachers play major role in selecting units
- Middle school mentor involved throughout
- Task Force teachers receive in-depth inservice on each unit
- Task Force teachers inservice and support other teachers at their grade level and school site
- Task Force teachers take responsibility for maintaining equpment kits

The product we were to develop was no less important than the process we would use to develop it. The process had to include the teachers in decision-making and result in strong, confident, school-based teams.

We began the project by surveying all K-5 teachers to determine the science lessons that were already being taught. A summary of this survey informed many of our decisions. This is where we learned that teachers relied primarily on textbooks, and that very few activities were being presented.

The first year of the project was aimed at selecting and testing eight science units — four for the fourth grade, and four for the fifth grade. Thus, we selected for the Task Force two teachers from each elementary school — one from each of these grade levels.

At the first meeting of the Task Force, Steve Gair outlined the overall goals and objectives of the program, and the curriculum specialists presented sample workshops using a guided discovery approach. The day ended with an overview of the forward-looking California Science Framework, which emphasizes hands-on science activities (at least 40%), and the importance of connecting units through big ideas in science (themes).

During the second session, the Lawrence Hall of Science staff presented a range of science units that supported the goals of the California Science Framework, and that were fully developed and classroom-tested. The Task Force teachers selected the most compelling units and determined which would best serve fourth-grade students and fifth-grade students.

Once these decisions were made, the Pittsburg Unified School District assembled kits for teaching each of these units, duplicating a set of kits for each school. Then followed a series of intensive full-day workshops on each of the selected units—one day per unit. Separate workshops were conducted for fourth- and fifth-grade teachers. Science curriculum leaders actually did the activities so that they would become skillful and confident in presenting them to their students, as well as serving as resource teachers to their colleagues. At the end of the workshop, teachers analyzed the unit for the concepts, science process skills, themes and scientific attitudes that it promoted. Between workshops, teachers would present the units to their students and come to the next workshop with reactions and questions about the unit they just completed.

At the final meeting of the Task Force, the fourth- and fifth-grade teachers again met together as a group, to identify the most important science processes and themes that should be emphasized through the units at their grade levels; these would be the focus of assessment instruments. The rich discussion on this day enabled the teachers to establish the connections between the science units in a year and from one year to the next. Teachers could see the various ways these units could be linked to help students build their science thinking skills and construct an understanding of major themes in science. This important session made the difference between the naming of four separate science units and the conceptualization of a well-linked science curriculum.

Unlike the science units, which had been developed and thoroughly classroom-tested by a variety of science curriculum projects at the Lawrence Hall of Science, the assessment instruments had to be invented and tested in this context.

Now that the project is well into its second year, three major activities are being conducted:

1) **Test Assessment Instruments.** The Task Force teachers are participating in the development of assessment instruments by trial testing them as they present the science units to their students. They are keeping written records of how they would change the instruments, so that they can be improved before being presented to all of the teachers in the district.

2) **Modify and Introduce the Curriculum.** The new curriculum units have been introduced to all teachers at the fourth- and fifth-grade levels in all seven schools. Workshops for each grade level were conducted at the district office, separately for each grade level. The middle school mentor teacher was the key person on the training team. Teachers performed sample activities from each unit, became familiar with the kit materials, and began to rely on the resource teachers in their own schools who serve as members of the Task Force. In addition to being able to rely on the Task Force member at their grade level in their school, the middle school mentor is available for ongoing support of teachers at all the elementary schools. As new teachers enter the system, there is provision for ongoing training of these new

OVERVIEW: A 5-YEAR PROCESS			
	4th-5th Grades	2nd-3rd Grades	K-1st Grades
Year 1	Invent & test the curriculum		
Year 2	Modify & introduce curriculum Invent & test assessments	Invent & test the curriculum	
Year 3	Train principals in good science instruction Modify & introduce assessments	Modify & introduce curriculum Invent & test assessments	
Year 4	Long-term support	Train principals in good science instruction Modify & introduce assessments	Modify & introduce curriculum Invent & test assessments
Year 5	Long-term support	Long-term support	Train principals in good science instruction Modify & introduce assessments

teachers, using both Task Force teachers and the middle school mentor teacher.

3) **Design and Test the Second-Third Grade Curriculum.** A new Task Force has been formed that includes two different leaders from each school; one at the second-grade level, and one at the third grade. The new Task Force will follow the same procedure as the fourth-fifth grade Task Force: plan the curriculum; hold workshops

for leaders; modify and introduce the curriculum to other teachers; and invent, test and improve assessment instruments for each unit.

Next year, the third year of the program, we will institute a two-day training program for principals. This workshop will focus on the new core science program for the fourth and fifth grades. It will help principals understand the goals of the science curriculum, how it relates to the State Science Framework, and in general, what good science instruction should look like. This will help principals support their teachers in their science teaching and inform them about how to evaluate good science teaching. Principals will be brought together to focus on the second- and third-grade science curriculum and the kindergarten and first-grade science curriculum when they are complete.

Also in the third year of the program, we will form a third Task Force, for K-1 teachers, and begin the cycle one more time. The K-1 leaders will introduce the curriculum to their colleagues and test assessment instruments in year four, and begin long-term implementation in year five. By the end of five years it is expected that there will be a team of at least five individuals at each elementary school, one at each level, who is an expert at each of the four units for that level, who will serve as a resource teacher for her or his colleagues, and who will maintain the kit of materials. At that time we will be able to determine whether we succeeded in achieving our goal—to design and implement a modern, up-to-date, activity-based thematic K-5 core science curriculum for all elementary schools in Pittsburg, California.

The Making of GEMS: Partners in Developing Curriculum

BY JACQUELINE BARBER

Great Explorations in Math and Science (GEMS) brings exciting and effective science and mathematics activities developed at the Lawrence Hall of Science into classrooms nationwide. These stand-alone curriculum units reflect the best facets of the "guided discovery" approach to science and mathematics education and are distinguished from many other published activities because they are assured to work! Far more than just a bunch of "neat" activities thought up by people in offices or ivory towers, GEMS activities are tested by hundreds of teachers nationwide, and modified according to their comments.

Ensuring that an activity will work in both Alaska and New Jersey, that it will work with 35 urban fifth-graders, in a carpeted classroom with no sink, even with a teacher who is intimidated by science — that is no small proposition! Extensive and varied teacher input into the curriculum development process is needed to make sure that activities work. GEMS activities have survived multiple rounds of scrutiny and critical use by a diverse group of teachers with hundreds of children. They represent a true collaboration between teachers and university curriculum developers.

A tremendous amount of planning and adjustment goes into the making of good curriculum units. Selection of a topic or concept to teach; original ideas for activities that involve students in learning about those ideas; designing those activities so they follow a cycle of learning, and thorough classroom testing in a wide range of situations. Likewise, a tremendous number of people contribute to good curriculum. GEMS is

Mining and Refining GEMS

1. Mine the Mother Lode.

Classes and activities developed at the Lawrence Hall of Science over the past 20 years are reviewed to identify those that are popular with students and communicate pivotal science and mathematics concepts and processes.

2. Select Ore Samples.

Reduce the pool of activities by selecting those that: a) rely on simple, inexpensive materials; b) are likely to work well in typical classroom situations; c) require no specialized knowledge to present; and d) complement other GEMS units to provide teachers with a full menu of choices.

3. Conduct the Acid Test.

GEMS author-developers test potential activities in local classrooms under the close scrutiny of the teacher and another curriculum developer. Activities that do not work well in the classroom are either modified or eliminated.

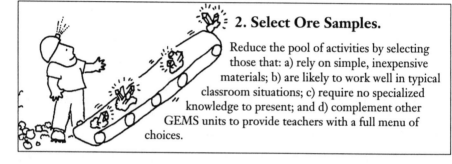

4. Rough Cut Gemstones.

Revisions suggested by the early classroom tests are incorporated into the first drafts of the GEMS teacher's guides. A committee of experienced curriculum developers and teachers examines and revises the drafts so they are clear and brief, yet contain all details needed to teach the units.

5. Try Out in Different Settings.

Twenty-five local teachers are given guides and invited to teach the units. These local field trials represent a variety of settings: urban and rural schools; experienced and novice teachers, and a wide range of age levels. Sessions are observed and recorded by a GEMS team member and teachers fill out detailed evaluation forms.

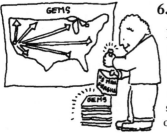

6. Fine Cut for Wide Variety of Settings.

Data from local trials are incorporated into revised teacher's guides and sent to 25 teachers at five national trial sites. Site directors conduct introductory workshops and gather evaluation forms from teachers for all sessions of each unit, and these are sent to GEMS Headquarters at the Lawrence Hall of Science.

7. Polish and Mount Completed GEMS.

GEMS author-developers summarize data from national trials, and complete final revisions of teacher's guides. These are then edited, illustrated, and printed in limited quantities, allowing for additional revisions as more teachers in the field begin to use them.

8. Introduce Completed GEMS Nationwide.

Dissemination of completed GEMS units to schools across the nation is in the hands of GEMS leaders, experienced teachers and educators who view GEMS guides as useful tools for teaching activity-based science and mathematics in their areas. Leaders conduct workshops for teachers and administrators, provide support for teachers introduced to GEMS materials, and keep in touch with GEMS users nationwide through meetings and the GEMS Network News.

the product of an enormous team of people: curriculum developers, authors, teachers, students, editors, artists, scientists and mathematicians, not to mention the numerous parents, scout leaders, hardware store clerks, rocketry salespeople, earthworms and other critters who have contributed to GEMS. Each plays an important role.

The cartoon metaphor on the previous pages provides a glimpse into the long and careful process by which GEMS are created.

Here are a few practical anecdotes to help illustrate the power of this curriculum development model. They are only the "tip of the iceberg" as the lessons and educational insight gained through such a process are numerous and always deepening.

In the GEMS Discovering Density activity, students layer different colored liquids in a clear drinking straw. A base of modeling clay was suggested as a way to both plug one end of the straw and hold it upright. However, sitting in the back of a particularly lively fifth-grade classroom, GEMS author-developers watched as children became obsessed with the modeling clay. Certainly there are many valid discoveries to be made with modeling clay, but in this case it was a serious distraction from the liquid exploration activities. This was confirmed by the many reports from local trial teachers, who spoke of the management difficulties they had with this particular session. Author-developers enlisted the assistance of this same lively fifth-grade class, telling them of the problem, and asking how it could be solved. "Don't use modeling clay!" they exclaimed. "It's irresistible! We can't help playing with it." What followed was a barrage of superb student suggestions for alternate materials, including our final choice — slices of raw potato.

Evaporation is a variable phenomenon, as we discovered through the national trial test phase. A liquid test activity that spanned two days needed rethinking, as the liquids evaporated overnight in dry Phoenix. Another activity, in which evaporation was important to the outcome of the experiment, proved to be problematic in Louisville. After two weeks, not only had complete evaporation not taken place, but mold had begun to grow, significantly altering the outcome of the experiment, to put it mildly. Teachers at these sites had a range of good suggestions for dealing with these variations, and GEMS author took these activities "back to the drawing board."

Product availability and variability constitute another potential pitfall of successful science and mathematics activities. The national trial arena helps us to realize and answer such questions as whether red cabbage is available year round in all places, the availability of the small red variety of earthworm, variations in the nature of tap water, and more. The information we get from these trials helps us know when to list alternative products, suggest unusual places to find them, or necessitate a major change in the activity.

Other lessons are larger. Some testing results provide a revealing window on the differences between a science center setting and a school classroom. Almost invariably, hands-on activities that might take a half-hour in the context of a science center class will take at least twice that (or longer) in a school classroom. To some extent, this is due to the larger class sizes in the typical classroom and less familiarity of the teacher with the material to be presented. However, just as often, teachers are more "in touch" with an appropriate learning pace for their students and are better at knowing how to pull all the learning opportunities out of an experience. This is particularly so when it comes to bringing an integrated discipline approach to activity units. A multiple-subject classroom teacher is a wizard at making literature and language arts connections in science and mathematics activities.

Each series of activities has led to its own discoveries during the testing process. Feedback is carefully analyzed. Many of the most appealing and imaginative notes and sidebars in the text come from this vast wealth of teacher and student experience and criticism. They work in the classroom because they were hammered out in the classroom. They appeal to teachers because they were shaped by teachers. This process of curriculum development, based on the strong collaboration between teachers and university curriculum developers, helps explain why when teachers pick up one of these guides they recognize almost immediately and invariably that it speaks their language and will "work."

Dr. Spencer Yost, a UCSF anesthesiologist, watches Mission High School student Luis Cabrera as he performs a DNA splicing experiment. Cabrera was among 280 biology students learning the latest in DNA technology as part of a university-school partnership (photo courtesy of UCSF).

(Right) Student winners display their trophies from the sixth annual Science and Health Lesson Plan Contest (photo courtesy of UCSF).

STUDENT CENTERED ACTIVITIES

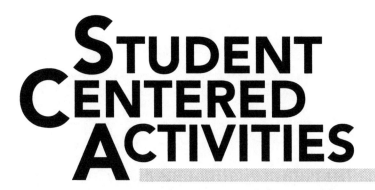

ATTRACTING UNDERREPRESENTED
PRECOLLEGE STUDENTS TO MATHEMATICS
AND SCIENCE DISCIPLINES

ONE SHOT ACTIVITIES:
SPEAKERS'S BUREAU AND SEMINAR SERIES

STUDENT OWNERSHIP OF SCIENTISTS' VISITS

PARTNERSHIP ACTIVITIES
OF HEALTH PROFESSIONAL STUDENTS

STUDENT LESSON PLAN CONTEST

SCIENCE AND SOCIETAL ISSUES SYMPOSIUM

SUMMER RESEARCH OPPORTUNITIES
FOR STUDENTS AND TEACHERS

ATTRACTING UNDERREPRESENTED PRECOLLEGE STUDENTS TO MATHEMATICS AND SCIENCE DISCIPLINES

BY SHELBY GIVENS

Attracting underrepresented precollege students to science and mathematics presents many challenges. First, many of these students, particularly those within inner cities, often have a sense of alienation regarding these disciplines. They know very few people like themselves who pursue these subjects en route to a career in mathematics and science-based disciplines. Next, many have been instilled with the idea (often by their teachers and even among their friends) that certain ethnic groups have more "ability" than others to do well in these subjects. In addition, many of these students are also pressured by their peers with the notion that to pursue school seriously and to attempt to excel in their subjects is tantamount to "acting white." There is a strong bias that only nerds are interested in mathematics and science.

This situation has very deleterious consequences for the individual students and for society as a whole. Technical literacy is a necessary key to many attractive and rewarding careers. In our highly technological society, science and mathematics illiteracy places the individual in a position of weakness and serious disadvantage. Currently, African Americans constitute more than 12% of the population, yet only 1.6%

of the nation's doctoral scientists and engineers are African American. Projections of current trends lead to a dismal picture of a predominantly white, educated elite pursuing rewarding careers while minority groups and uneducated whites are relegated to the poorly paying service sector.

Changing demographics emphasize the urgency of our task. Projections indicate that today's minorities will constitute a majority of the nation's population before the end of the next century. That reality has already impacted many of our schools. More than 30% of today's public school student body is minority. By the time you read this, California and Texas, two of the most influential states in precollege education, may have joined New Mexico and Mississippi in having "majority minority" public school systems.

Given this situation, efforts must be made to change the attitudes of underrepresented students and to link them with mathematics and science teachers who have high expectations for all their students. The groups that are most commonly historically underrepresented in mathematics and science include African Americans, Native American Indians, Mexican Americans and other Latinos. They provide a unique example of groups of students that could benefit from a shared awareness of the unique contributions that their cultural forebears have provided toward humanity's present level of comprehension of the mathematics and scientific disciplines. Both Africa and the Americas featured highly developed systems of engineering, astronomy and medicine. Using these examples students can realize that their interest and potential mastery of these disciplines is indeed part of an ancestral tradition and that their participation within these subjects is consistent with who they are. Illustrative posters, books and videos that depict these themes would be appropriate and useful in reshaping student attitudes toward these subjects.

In addition to historical examples, many African Americans, Latinos and Native American Indians currently succeed in highly technical fields. Many of them had to overcome enormous obstacles without significant support. Fortunately, many programs in a variety of geographical regions and cultural perspectives now exist that effectively help underrepresented students to succeed in pursuing scientific and mathematics disciplines in school and eventually in the professional work-place. Unfortunately, we still have a long way to go in committing the leadership, staffing and funding required to serve the large number of students that are ready for and need these programs.

The California Mathematics, Engineering and Science Achievement (MESA) program is one of the success stories. MESA began in the late 1960s when a Berkeley engineering professor and an Oakland high school mathematics teacher collaborated to increase the alarmingly low number of black students who entered the university's engineering program. They reasoned that if teachers worked with promising student scholars the same way that coaches help young people who demonstrate athletic prowess, then Oakland could become as famous for its minority engineers as it is for its athletes. From this beginning, MESA has grown to currently serve approximately 15,000 students in California's schools.

> **If teachers worked with promising student scholars the same way that coaches help young people who demonstrate athletic prowess, then Oakland could become as famous for its minority engineers as it is for its athletes.**

The program prepares historically underrepresented students from elementary through university level for careers in science, engineering and other math-based fields. Over 80% of all MESA high school graduates go on to a four-year college or university following graduation, and 65% of them major in science, engineering or other math-based fields. Students who participated in MESA's Minority Engineering Program accounted for two-thirds of the minorities who received bachelor's degrees in engineering in California in 1990.

An effective MESA precollege program involves a primary partnership between a MESA center and a school district that results in the commitment of human and financial resources, a dedication to academic success for MESA's targeted students, and a diligent effort by all partners toward realizing these aims. Generally, a MESA precollege program center is housed within a school of engineering or science on a university campus. Every MESA center is headed by a center director who is responsible for delivering a range of support services to students and their advisors at selected elementary, junior and senior high schools. MESA advisors are usually mathematics and science teachers.

Program activities include several components that help to create a sense of community among underrepresented students in mathematics and science disciplines. The program may include the following components:

1. Organized study groups and tutoring by role models.

2. Academic enrichment experiences including innovative curricula, field trips, chapter meetings during and after school, and summer sessions.

3. Career exploration including presentations by role models, counseling regarding necessary course work, and field trips to work sites.

4. Early identification, intervention tutoring, academic counseling, follow-up, and career exploration with an emphasis on utilizing role models.

5. Incentive awards including scholarships for secondary and undergraduate studies.

6. Organized parental involvement in all program activities.

7. Leadership development and peer support.

At middle schools and high schools, MESA students are organized as a chapter or club that has a "MESA period" that regularly meets under the guidance of a mathematics or science instructor designated as the MESA advisor. This "MESA period" organization enables the clustering of students who share a common interest in math and science, and minimizes the adverse influence of students who may ordinarily be in a math or science class only because they "have to." During the "MESA period" class, students are surrounded by peers for whom it is normative behavior to be interested in math and science. The "MESA period" allows students to work in a collaborative manner on science projects that they may be developing for subsequent competition between themselves and other MESA students from different schools. An annual event called "MESA Day" provides the occasion for MESA students from a variety of schools to come to one university location and engage in a variety of academic competitions. In this setting they are able to encounter hundreds of students who look like themselves and have a serious interest in mathematics and science. This experience further normalizes for them the acceptability of excelling in these subjects. As a guiding principle, MESA sets a standard of high expectations for success in mathematics and science.

As an addition to the "MESA period," a Saturday Academy may be instituted to serve a group of underrepresented students. The course offerings may include mathematics, science, and verbal and written communication. The math and science should stress hands-on activities that apply concepts relevant to the students. A Saturday Academy or a summer program provides another context for students to come together from different school sites. This situation avoids much of the distraction that normally occurs during the regular school day. Also, those students who take math and science classes because they "have to" will generally not attend sessions outside of regular school hours.

In general, the MESA approach to excite underrepresented students about mathematics and science involves an emphasis on hands-on activities. Further, these activities are most effective when they directly relate to the students' everyday world. Every effort should be made to heighten students' curiosity about things in their immediate environment (home, neighborhood or community). Those things which the student takes for granted can be related to general scientific principles, e.g., gravity, electromagnetism, solar energy. Attempts to answer questions such as "how does a steel ship float and yet a small nail sinks" can introduce the concept of buoyancy. This concept can be explored using "foil floaters" that are fashioned as a flotation device that is then filled with as many pennies as possible before it sinks in a container of water.

Every effort should be made to utilize play or recreational activities as a basis for generating interest and curiosity about the subject in question. Middle and high school students enjoy paper airplanes especially if they are encouraged to design their own and then compete to maximize "hang time" or distance traveled. This would be an excellent time to invite an aerospace engineer to discuss with students the principles of aerodynamics that affect the relative successes and failures of aircraft design, as well as give a personal perspective on technical careers.

Once students become curious, they are eager to use books as additional references or resources for further research. They are motivated to seek information that they truly want. This is a very different and much more effective approach than first exposing students to the texts and making them absorb information via rote memorization.

Since MESA includes underrepresented students from elementary grades through its university level engineering program (MEP), we have a pool of older students who can serve as role models for younger students. MEP students on university campuses often serve as tutors for MESA secondary students at the high school level. High school students often serve as tutors and role models for middle school students. Since MESA follows the activities of its alumni professionals, these engineers and scientists are also utilized as speakers and role models. By cutting across age differences, students at all levels gain a clear sense that continuity of their best efforts can lead to success in mathematics and science-based careers.

The majority of MESA students come from homes where parents or guardians have little formal higher education. When a MESA student succeeds in attending college, he or she frequently is the first in the immediate family to do so. Despite the many obstacles that currently deter underrepresented minorities from pursuing mathematics and science-based disciplines, we have demonstrated various measures that actually change student attitudes and that result in success in these fields. Students discover that science and math can be fun, that these fields are related directly to their lives, that their cultural history includes significant accomplishments, and that people just like them are interested in these disciplines and have successfully pursued satisfying technically skilled careers.

Many of the approaches that enhance the achievements of underrepresented groups can be brought into the school system to benefit everyone. Educational excellence and equity can proceed hand in hand.

ONE-SHOT ACTIVITIES: SPEAKERS' BUREAU AND SEMINAR SERIES

BY DAVID STATES AND MARGARET CLARK

Many institutions provide access to a speakers' bureau, contest judges, or tours of facilities to teachers or schools that request them. Since these activities rarely involve repeated contact between individual teachers and scientists, we call them "one-shots." They do not tend to build the kind of strong personal or institutional bonds that develop with the more intensive one-on-one partnerships. Although we believe that it is a mistake to place one-shot activities at the center of any outreach program, the availability of a pool of volunteers that will fulfill one-time requests can be a very effective component of your program.

There are three major reasons for having some kind of one-shot program. First, there is a strong demand from teachers and school administrators for outside experts to fill gaps in the curriculum or to provide role modeling for students.

Second, there will always be a significant number of potential participants at schools or a university whose time is so limited that they are reluctant or unable to make a sustained commitment to the program. This problem is particularly true for senior faculty at the university. Because they can provide powerful support, recognition and enthusiasm for the program, it is important to develop avenues for these prestigious individuals to participate. Many teachers are also too busy to commit to a continuing interaction with university volunteers, and prefer an activity that requires only limited commitment.

A third reason for organizing one-shot activities is that they often serve as a low-risk introduction to the program for volunteers and teachers who hesitate about participating. Once they have an initial experience of the program, many of these one-shot participants enthusiastically come back for more and deeper involvement.

SEP maintains a database listing approximately 300 university employees who have volunteered to go to schools one to three times a year to give talks on specified topics. Topics listed by speakers cover a wide range, including careers, basic science issues, and others relating to health and disease (see listing of selected lecture topics). The speakers' bureau consists principally, although not exclusively, of faculty members. We can thus utilize their prestige and expertise by offering them one of the few activities that fits into their extremely busy schedules. This is also a good way to utilize the limited number of minority scientists, who, along with women, are particularly in demand. Titles for available talks are published and distributed to teachers once a year. In addition, we make an effort to fill requests on any topic by cross-referencing subject listings kept on our database and asking faculty members whether they would be willing to address that particular subject. Most faculty, with a little encouragement, will provide talks even if they fall somewhat outside their specialized area of research. In fact, some prefer this opportunity, and these talks may be easier for them to give at an appropriate level.

Over the years, we have narrowed the list of speakers by eliminating those volunteers who lack charisma or the ability to adjust from a college-level audience to one consisting of high school or middle school students. Although we prefer to use well-known experts in the field, it is more effective to use a graduate student who communicates well rather than the internationally famous professor who bores even his colleagues. On the other hand, we have Nobel laureates who are also compelling speakers. Through experience and research, one can build an excellent list of topics that are delivered by strong speakers.

We have shortened and simplified our lecture list to make it more applicable to secondary education needs. This list is extremely valuable in providing ideas and inspiration to the teachers and their students. Since a long list of esoteric talks is likely to discourage teachers from

Sample Topics

The Brain: How the Brain Works
The Brain in Youth and Old Age
Language and the Brain
Memory and Learning
Vision, Perception and Optical Illusions
Pain
How Drugs Affect the Brain

Biochemistry/Cell Biology
Development
How Cells Grow and Divide
How Cells Talk to Each Other
The Skeleton of Cells

Cancer
Misguided Cells: The Genesis of Human Cancer
Detection and Prevention of Cancer
Treatment of Cancer
Cancer and DNA

Careers
Careers in Dentistry
Careers in Medicine
Careers in Nursing
Careers in Research Science

Diseases
AIDS
Allergies and Autoimmune Diseases
Epidemiology: The Detective Process
Genetic Diseases
How Our Bodies Fight Infection
Immigration and Health

Drugs and Alcohol
Advertising — Selling Tobacco and Alcohol
Drugs, Alcohol and Pregnancy
Good Drugs, Bad Drugs
Second-Hand Smoke

Healthy Lifestyles
Cross-Cultural Approaches to Medicine & Nutrition
Healthy Lifestyles and At-Risk Behaviors
Preventive Medicine
Sports Medicine

Human Reproduction
Birth Control
Prenatal Diagnosis and Care
Pubertal Maturation
Sexually Transmitted Diseases

Molecular Biology
Biotechnology and Recombinant DNA
Computer Modeling of Molecules
DNA for High School
Human Genome Project
Molecular Evolution

Nutrition
Biochemistry and Dieting
Eating Disorders
Four Food Groups: Is It a Hoax?
What Your Body Does with Food

Scientific Process
Animal Research — Is It Necessary?
Ethics in Medical Practice
Ethics in Scientific Research
Scientific Method and Philosophy

requesting speakers, be selective in your listing and use titles that convey ideas that are meaningful to the students.

A one-shot speaker program for schools has several attractive features. First, organizing talks is straightforward. Moreover, once a list of interesting and reliable speakers is established, responsibility for facilitating the assignments can be delegated by the coordinator to an administrative assistant. From the teacher's perspective, outside speakers, especially scientists or physicians, enhance the teacher's credibility with the students. More subtly, their presence indicates that they think teaching students about science is important. Equally important is the effect on the scientists of visiting schools. They have a chance to meet students and learn about their interests, and they usually develop a greater appreciation for the work teachers are doing.

A speakers' bureau also has both avoidable and unavoidable limitations. Unless one takes care to select good speakers and topics, and to provide sufficient background to the speaker on the expected audience and their level of previous training, speakers can easily lose the audience and put students to sleep. Even high school students have little tolerance for a straight lecture format. Novice speakers are often astounded by the low level of scientific knowledge displayed by 15-year-olds. For these reasons, we require all speakers to discuss the audience beforehand with the requesting teacher and strongly encourage the inclusion of either hands-on demonstrations and/or informal discussions with students. Many potential lecturers have unrealistic expectations and need to be convinced that how they communicate is at least as important as the content of their communication.

It is often difficult to coordinate the scheduling and content of one-shot lectures with the teacher's curriculum plans. Even lectures on important topics can be lost in the shuffle if they seem unconnected to the students' lives or what they are learning in class. Another practical problem is that a volunteer lecturer may have time to talk with only one of two of a teacher's classes, but the teacher would like all her students to hear the speaker and also prefers to have all her classes remain on the same daily schedule. Because of the limitations of one-shot lectures, we think a speakers' program must be viewed as an enhancement tool that can provide additional interest and information, rather than the major component of a science education outreach program.

The North Carolina Museum of Life and Science wrote and distributes an eight-page guide entitled "Sharing Science with Children: A Survival Guide for Scientists and Engineers." To receive a free copy, write to their Director of Education at P.O. Box 15190; Durham, NC 27704 and include a brief statement explaining how you plan to use the guide. We have included below some of the tips from their guide.

- **Know when and where you will be visiting.**

  Verify the time, place and length of the visit. Be sure to get phone numbers for the teacher and the school. If you don't know where the school and classroom are, ask for directions.

- **Share yourself.**

  Let the children know you are a real person with family, pets, hobbies. Talk about how you got to be chemist, an anthropologist, an engineer.... Was there a special event or person in your life — a teacher, a learning experience, a book, a visit to a museum — that aroused your interest in your field? What do you do on an average day?

- **Involve the students in doing.**

  Bring an attention grabber if you can. Keep in mind that your goal is to arouse curiosity, excitement, eagerness to know more. The tools of your profession may be commonplace to you, but they are mysterious, unknown, even fascinating to most of the students (and teachers) you meet. When possible, let students handle models, equipment, samples, plants, prisms, stethoscopes, rocks, or fossils. Have the students participate in using the processes of science — observing, identifying, classifying, measuring, hypothesizing.

- **Stimulate thinking by asking questions.**

  Questions that ask students to make a prediction, to give an explanation, to state an opinion, or to draw a conclusion are especially valuable. Be sure to allow time for each student to THINK before anyone gives answers.

- **Use language the students will understand.**

  Be conscious of vocabulary. Try not to use a difficult word when a simple one will do. Define words students may not know. For example, don't say, "I am a cytologist" and begin a lecture on semipermeable cell walls. Rather, ask students if they know what a cell is and then tell them you study cells, how they are built, and that you are called a cytologist.

- **Make what you are talking about real to the students.**

  Show the students that the area of science or technology you work with every day is part of their everyday lives, too. How has what you and your colleagues have learned up to this time changed how we do things or understand things? How will what you do make students' lives better or different in the future? How does what you do and know relate to what they are learning in school?

Over the years, we have experimented with ways to maximize the usefulness of one-shot speakers. For several years we organized a monthly seminar series to update teachers on central themes in biology. It is difficult for teachers to keep up with recent advances in science and health professions. A subcommittee of the steering committee (teachers and researchers) determined the initial topics, and posters were sent out in advance to each school inviting teachers to attend. UCSF scientists were briefed ahead of time by teachers on district curriculum related to the chosen topic, and their talks, which updated teachers on basic theory and current perspectives, were very informal and featured many interruptions with questions and answers.

A secondary, albeit intentional, effect of these "Curriculum Updates" was to create a regular forum for the district's science and health teachers to meet with one another, discuss mutual concerns, and plan strategies for addressing their professional problems. Under normal circumstances, these district-wide meetings of science or health faculty had been rare or nonexistent.

After two years, attendance at the Curriculum Updates decreased significantly. Having covered many areas of general interest (proteins, DNA, viruses), the talks began to focus on more specialized topics and attract an enthusiastic but very limited audience. SEP coordinators and the steering committee eventually decided to discontinue the Curriculum Updates in order to make more efficient use of speakers and program coordinators. We think it is important for a program to have the flexibility to adjust to changing realities.

To continue using volunteer speakers in an effective manner, we organized monthly and biweekly seminar series for students at individual high schools. In this program, teachers and/or students from each school choose the format and time for the talks, define ground rules for attendance, and select topics from a menu provided by the SEP coordinators. It is extremely important that the school provides input into the series and takes some responsibility for its success. SEP finds an appropriate speaker, schedules the talk, and prepares eye-catching posters. As with all one-shot speakers, we take pains to preselect and prepare speakers in order to maximize the effectiveness of their talks. Although the format varies from school to school, the best series tend to be ones that emphasize a thematic approach to the talks. For example, one school had several talks on different aspects of the brain and arranged the lectures to follow one another in a logical sequence.

Some of the seminar series are organized as science club meetings. Other schools have chosen the option of having talks during class time and allowing students from various classes to attend if they meet certain criteria, such as keeping up with their work. Assembly hall presentations are discouraged because it is rare that a speaker will have the charisma to maintain the attention of several hundred students. Further, mandatory attendance seems only to spread the virus of boredom from uninterested students to those who normally would be engaged. We keep the presentations small (20-40 students), and, as much as possible, we try to heighten the profile of the seminars, present them as special occasions, and institutionalize them in the process.

The seminar series has the advantage of converting a one-shot volunteer activity into a longer term and more effective institutional link between the university and the schools. The SEP Science and Health Lesson Plan Contest (see article on p.157) provides another mechanism for establishing an important program that utilizes limited volunteer commitments. In general, program coordinators should enhance one-shot activities by utilizing them within a more sustained and broader framework.

STUDENT OWNERSHIP OF SCIENTISTS' VISITS

BY DEAN MULLER

Last fall, I received an impressive list of possible speakers and topics for one-shot presentations. At first, it was an overwhelming embarrassment of riches. How could I possibly integrate these authoritative and technical speakers and topics into my one-man science department in our small continuation high school?

Well, when in doubt, I sought help from that endless well of support, the students. I created a process whereby the students could work in small groups to discuss possible talks that they would like to hear. Students formed groups of four to six individuals. Each group selected a leader who made sure that everyone had a chance to participate. The group also selected a recorder to keep notes of the discussion and conclusions. I even created a form for the groups to record the names of the participants and their conclusions.

This activity provoked a lot of discussion concerning the topics, especially those with titles that were somewhat technical. The students learned the meaning of new words and started to get a sense of the division of labor in scientific research fields. The students took the activity seriously since they would decide who the speakers would be.

I read through all the responses and got a sense of the clusters of interest and priorities of the students. Then I made copies of all the group papers and distributed them. Not surprisingly, sex and drugs topped the list of topics on the minds of students. Other areas of interest included pregnancy, heredity, cancer and stress.

Working with SEP Coordinator David States, we quickly arranged for the first speaker, Dr. Li Bero. Her area of expertise is the effect of drugs on the human body. Dr. Bero and I arranged a date and time for her presentation. A little over a week before her scheduled visit, I reminded students of our project to bring in UCSF speakers and told them about Dr. Bero. My purpose was to create a process and a context for students to take a responsible role in their education. With that made explicit, the students worked in small groups to generate a list of questions for the speaker. They were instructed that this exercise was real. Their previous responses had led to the selection of talks and their questions would now be given to the upcoming speaker.

Dr. Bero was wonderfully helpful. By the time she arrived she was very aware of her audience and their interests. When the students realized that their input had been taken into account, their enthusiasm for these presentations increased.

In the continuation school setting I use a variety of techniques to try to get maximum student participation. I remind students each day for a week prior to the talk. Every day of the semester I have announcements and instructions taped to every table in the room. I include a Gary Larson cartoon to catch their attention. On these announcement sheets I invite the students from all my classes to get a pass from me so that they may attend the talk. The presentations are usually scheduled during my fourth-period class as that is the time when the greatest number of students are in the school.

This process worked extremely well. The students who attended were interested and attentive. They would get involved with the speakers and ask questions. Students would often stay after the presentation to talk with the speaker. As the students became more familiar with this procedure, they would ask me when the next speaker was coming. They looked forward to these presentations and asked if we could have them more often. If anything, the changes I would make would be to give students even more control over making arrangements with speakers and perhaps begin to design follow-up activities or discussions.

Partnership Activities of Health Professional Students

The Experience of the Health Education Partnership at U.C., San Francisco

BY JOANNE MILLER, GREGORY ARENT AND SARAH SPENCE

Background

The University of California, San Francisco (USCF) is the only campus of the UC system that has no undergraduates. It is a Medical Center that features a very large research and graduate education program in the life and health sciences, a medical school, a dentistry school, a pharmacy school, a graduate nursing school, and various specialized institutes. In 1988, the Science Education Partnership (SEP) program, a partnership between UCSF and the San Francisco Unified School District (SFUSD), was begun by Dr. Bruce Alberts, chairman of the Biochemistry Department. After this UCSF precollege science education outreach program had operated for a year, it became apparent that the university had as much to offer in the field of health education and the school district had even greater needs in this area.

While looking for funding to expand the science education program into the health sciences, the university learned about a "Comprehensive School Health Education" grant program of the U.S. Department of Education (DOE) Fund for Innovation in Education. UCSF engaged JoAnne Miller, M.A., a veteran teacher and former UCSF health educator, to assess existing campus health science volunteer activity and to apply for the DOE grant. The major three-year DOE grant for the Health Education Partnership was awarded in October 1989 to create site-specific health education programs at targeted middle schools.

This program was housed with SEP and JoAnne Miller was retained as director to coordinate the Health Education Partnership (HEP) activities of SEP. This partnership consists of various programs and interactions between the UCSF staff, faculty and students, and the students and staff of SFUSD. It is dedicated to reaching under-represented, at-risk middle school students at seven school sites by role and career models who deliver hands-on health science experiences and information on risk-taking behavior.

The program focuses on middle schools because adolescence is a critical period in which behaviorial patterns are set that most likely will last a lifetime. School district administrators targeted middle schools because of the district's lack of health education curriculum in grades 6-8 and their recognition of the urgent need for health information for this crucial age group.

Most SFUSD youngsters are considered at-risk and even high risk because of their socioeconomic status, and high percentage of ethnic minorities and recent immigrants. Each year, one of every 10 students in the SFUSD is attending school in the United States for the first time. Approximately 30% of the students have a Limited English Proficiency. These students clearly need health-related information to make informed decisions.

The program began by the school district and HEP inviting middle schools to participate in the program. The premise of the HEP program's relationship with the individual middle schools is that the student needs for health information would drive the type of resources provided to each school by UCSF. An annual conference is held with the school site committee, composed of the administration, teachers and parents (wherever possible), and the HEP staff to plan the partnership activities. The committee assesses what worked, what was not as productive, and what health-related challenges the in-coming students will bring. The program is then finalized for the school year.

The committees all identified great needs for instruction in the areas of maturation, substance abuse, preventative health and sexuality. It became clear that UCSF faculty volunteers would not have time to equitably serve all of the school sites. A bigger and renewable pool of volunteers needed to be developed. Further, because of the special needs of many of these these youngsters, it became clear that they must be reached by sensitive, role models who are closer in age and culture to them than the current teaching force.

The HEP director realized that a good solution would be to enlist the UCSF health professional students. In the first quarter of HEP's operation, two medical students volunteered so successfully on a weekly basis in a science classroom that the HEP director sought out the deans of the individual schools to see how such opportunities for consistent volunteer work could become part of the regular university offerings. The HEP director created two elective credit courses, one each in the School of Dentistry and in the School of Medicine, that have provided the focus for recruitment of health professional students and for the health education outreach.

Students are recruited for these courses and other HEP activities at the annual Student Orientation Fair put on by UCSF in the first week of each new year. They are also recruited by current student volunteers who make presentations to classes each quarter.

Dental Public Health 188

Dr. John Greene, dean of the School of Dentistry, responded immediately by putting HEP in touch with Dental Public Health professor, Dr. Howard Pollick. The HEP director, under Dr. Pollick's guidance, instituted a new course, Dental Public Health 188, with Dr. Pollick as the instructor, dedicated to bringing dental health science education to the middle schools. The first quarter offering was in the spring of 1990. It has continued to be offered each quarter with a steady core of enrollment.

Dental Public Health 188, the vehicle for dental and dental hygiene students to volunteer to work in the schools, is, in the students' words: "This is the only time we can get out of the lecture hall or lab and have real hands-on experiences. Otherwise, we'd never get to go near a patient until the third year." "It really gives us a boost up on the other students because we have had numerous experiences doing screenings and just getting close to people." "Most importantly, it gives us an opportunity to teach what we have just learned. This class pushes us out into the community to work together, and gives us a real feel for public health work."

The HEP director attends, plans and co-leads the weekly class with Dr. Pollick. She is responsible for the coordinating and monitoring of students with the school site. She instructs students in appropriate curriculum and teaching skills and supplies students with resource materials. Students borrow models, slides and other equipment from the Resource Center at the SEP office. The class and Dr. Pollick advised which materials to purchase.

Although each quarter the focus changes slightly, all students deliver dental health and hygiene information in unique, hands-on methods to middle school youth. Dental students have developed a skit — including costumes — about bacteria in the mouth that the middle school students loved. Students in the elective class also research the dental education literature, and write reports suitable for publication or award by a professional association.

Working with the SEP science director, students put together and piloted a three-week, hands-on dental science curriculum based on the California Science Framework. Since most dental diseases result from the action of acid-forming bacteria, these instructional modules include lessons on pH and bacteriology as well as dental hygiene. Many groups approach science teachers to include their particular agenda in the already crowded science curriculum. Teachers naturally resist these intrusions. In contrast, science teachers responded enthusiastically to the dental lessons because the lessons incorporated topics that are already in the curriculum and because students are excited by lessons that relate to their bodies.

The dental students also put together resources for the SEP Resource Center such as X-ray sets, sets of molds of mouths and their own slides, to be made available to whomever requests items. They decide on how they put together the presentations and which materials they need.

MEDTEACH (EPIDEMIOLOGY 198)

MedTeach grew out of a small grant from the American Medical Student Association for first- and second-year medical students to work with at-risk populations in San Francisco's Youth Guidance Center (YGC) on substance abuse issues. At the close of that grant, the medical students wanted to continue their volunteer activities. The students approached the HEP director with the plan of becoming the medical-student arm of HEP. MedTeach has now become the most active part of HEP, as well as an accredited elective course.

Twice a week medical students teach a one-hour, health-related lesson at YGC. One morning each week, teams of medical students arrive at middle schools to teach expanded versions of these lessons. They provide important hands-on learning experiences and information regarding health and wellness, the human body, substance abuse and mental and physical maturation through weekly site visits. They serve as role models and include career education in their presentations.

> *The result is that UCSF trains doctors who are more understanding and culturally aware; and SFUSD has more highly motivated students who will seek healthier lifestyles and may be influenced to pursue careers in the health sciences.*

The HEP staff coordinates and monitors the student teams with the school site. The staff makes sure the teams are appropriately constituted for the school site and that lessons are age-appropriate. The staff is available to consult on curriculum and teaching techniques. HEP staff supply medical students with models, charts, real human anatomical specimens, and curricular materials. The HEP staff administers MedTeach volunteers whether they are applying for course credit or not.

The MedTeach model impacts a diverse student population with new curriculum that has been added to both the university and the schools. MedTeach enriches each institution and its respective students and bridges the gap between the teachers and the university, and the students and higher education. The result is that UCSF trains doctors who are more understanding and culturally aware; and SFUSD has more highly motivated students who will seek healthier lifestyles and may be influenced to pursue careers in the health sciences.

The basic methodology in the MedTeach plan is that teams of medical students aid teachers in the school district with their science curriculum, especially as it pertains to health-related issues. Volunteers are drawn from the first- and second-year medical school class at UCSF and teams of 3-4 students are assigned by the director of HEP to a given middle school.

The teams are expected to make contact with the 6th-grade science teachers at their assigned school to discuss the program and their thoughts about the project. At this point, the individual programs diverge according to the needs of the teacher and the school's student population. One of the beauties of the MedTeach program is that the methodology of the individual teaching teams is as varied as the personalities and goals of the members of the team, the many schools to which they go, the teachers with whom they work, and the students with whom they interact. Although the anatomy and physiology lessons presented to all of the 6th-grade classes cover the same information, placement and functions of body organs, instruction of how substance abuse and ill health affect each organ, and the handling of the different organs with gloved hands by all the students, MedTeach students are free to find their own means of expression.

As an illustration of the diversity and flexibility built into the MedTeach program we have included descriptions of a few experiences from different teams collected during the past year.

1. One group taught the 6th-grade class at Horace Mann School — a culturally diverse middle school in the Mission District in San Francisco. Since the teacher had spent quite a bit of time on normal physiology already, they decided to focus on the pathophysiology of drug and alcohol use. They met once with every 6th-grade class over the course of several months in separate one-hour sessions. During each visit they started by giving a general review of organ anatomy. This was accomplished by selecting a volunteer, instructing the person to lie on the floor and having another volunteer trace the student's body on a sheet of paper. After the outline was obtained, the

other students had to place colored paper cutouts of internal organs in the appropriate positions on the tracing — explaining what the organ was and its function while they did it. Next, each team member took a subset of the class and explained the effects of drug and alcohol abuse on a specific organ system of the body. Visual aids of human organ specimens were employed to give the students a more concrete sense of what abuse can really do: a cirrhotic liver to illustrate alcohol abuse; carcinomic lungs to illustrate smoking risks; and a human brain to show where drugs have their mind-altering effects. Latex gloves were supplied to allow them to touch the anatomical specimens. The team also covered the information "what is a drug?," and whether the students had ever been exposed to alcohol (an anonymous survey).

2. Another group followed the same class over an entire quarter on Wednesday mornings for about two hours each visit. The class was an English as a second language (ESL) group made up of sixth-graders at Marina Middle School in San Francisco. The group of four medical students started by introducing themselves and telling the class how they became interested in the medical profession. To determine the level that they should begin teaching, they played ice-breaker games with the class that introduced body parts and functions. The class then split into three smaller groups, rotating among three stations, allowing the students to learn on a more personal level. Stations in the first class included a human model, a collection of human bones, X-ray films, and a photographic atlas of the human body. Students were encouraged to touch and feel the models and bones and to ask questions. The teacher was interactive with the medical students and had the class make up questions about what they learned to be used in a "Jeopardy" game they played midway through the week.

The second week focused on the lungs and the liver. Actual human lungs and cirrhotic liver samples were used, along with microscopes and slides of healthy and damaged lungs. The medical students again divided the class into three smaller groups — the liver and alcohol, the healthy lung, and the damaged lung. The medical students taught the sixth-graders about the normal physiology of the different organs, and how they were affected by alcohol, drugs and smoking.

Students were able to hold a cirrhotic liver in their hands and actually feel the result of alcohol abuse, not just read about it in a book. They worked through several questions dealing with peer pressure to try drugs and alcohol, and the sad reality of family substance abuse and how to get them to quit. The class ended with the students making up questions for their new game.

In the third week, the groups discussed the heart, the circulatory system, and the adverse effects of smoking, bad diet and drugs. A human heart was used, as were plastic models, charts, microscopes and slides for the three discussion groups. The groups talked about family members who had heart attacks and what caused them, the benefits of exercise and other ways to decrease their chances, as well as their families' chances, of having a bad heart, and recent advances in medicine that led to the ability to transplant hearts. The class continued their practice of writing up questions to be used in their game.

The fourth week dealt with a class choice of topics—the brain and senses. Through UCSF's Department of Anatomy, the medical students were able to find a human brain and spinal cord. Along with the brain, other teaching devices were utilized such as microscopes and slides, the human skull, peripheral vision testing machines, as well as charts and models. The three groups discussed the brain and its functions, the senses and how they work, and problems that are caused by drugs and alcohol. Students were encouraged to get as much hands-on time with the brain and models as possible and to discuss the reasons people use drugs and alcohol.

3. Another volunteer worked primarily as a curricular consultant with the teachers in a middle school, developing detailed teaching plans for a variety of health- and science-related topics, including basic anatomy, physiology, and different drug effects on the body and health.

All of these different approaches were designed to meet the common goal of the program but were tailored to the specific needs of the schools in which they were implemented. We believe that this flexibility is one of the things that makes the MedTeach program so successful.

In its second year, the MedTeach program has reached approximately one-third of the students in the past two Medical School classes. This program has enabled students to:

- reinforce what they have learned in their university classes
- increase their sensitivity to the future patients they will serve
- train them to be teachers of preventive medicine
- integrate the community into their early training
- gain clinical experiences for them early on in their career training
- widen their horizons for future career opportunities by broadening their medical practice options
- help them realize that there is a large pool of medically underserved people for whom they share a responsibility.

The program has been valuable to the students and their interest and commitment to the project is mirrored in the fact that most of the students enrolled do not take the course for credit.

MedTeach began in 1990 with only 8-12 medical students and expanded in the second year to encompass over 40 medical students helping eight different SFUSD middle schools. The scale of the program is enormous — in 1991-92, medical students involved with the MedTeach program reached more than 6,500 students in San Francisco's public schools. The benefits of the program to medical students, teachers and, most of all, the adolescents involved are clear. The volunteers consistently report high levels of enjoyment and personal satisfaction with their work in the middle schools. They enjoy being able to apply the medical knowledge that they have recently acquired, as they do not normally have the opportunity to use their knowledge with patients until the third year of medical school. Responses are similar:

> "It was fantastic to be actually contributing something to the community instead of fixating ourselves only on our classes and scores and performance. You really are reminded of why you are in medical school in the first place when you work with kids like this."

> "It was hard work at times and demanded a full level of energy input on our part. But the rewards were tremendous — seeing them respond, hearing their eagerness."

> "Learning more about kids at this age, their interests and develop-

mental levels will certainly be helpful in my future work. I also learned some good interactive teaching methods that I think will be helpful."

MedTeach is also beneficial to the teachers involved. The medical students provide much-needed educational resources and are able to tailor their lesson plans to each individual teacher's and classroom's needs. This type of teamwork serves an instructive role for the teacher, providing suggestions and opportunities for varied teaching methods.

However, the most beneficial aspects of MedTeach are the effects that the program has on the adolescents. The ethnic diversity of the volunteers, their closeness in age to the students, the casualness of their dress and speech, and their nonjudgmental perspective toward teaching has a large impact on the students, and provide them with much-needed positive role models. The novel ways in which the students are able to learn — visual aids, role-play, small group discussion, activities and presentation of organ specimens — make learning a whole new and pleasant experience for them. Being able to hold a human heart or a human brain in their hands gives them new respect for their bodies. The fact that they can feel and inspect a diseased organ instead of simply hearing about it makes the risks of substance abuse more real to them. The program has even farther reaching effects given the fact that students are also interested in ways to help their friends and family members engage in health-related behaviors such as to stop smoking, drinking or abusing drugs. Many of the adolescents tell volunteers that they never really liked science before, but now are interested in entering the health science field.

THE STUDENT LESSON PLAN CONTEST

BY ART SUSSMAN

E ach spring, hundreds of San Francisco middle and high school students blossom into science teachers. The annual event that triggers this process is the Science and Health Student Lesson Plan Contest. When UCSF's Science and Health Education Partnership (SEP) had an extra $5,000 in its first year, Professor Stan Glantz suggested giving the money to students. His provocative suggestion resulted in the contest we have come to know and love.

The concept is simple — students work in teams to develop a lesson that they teach for one class period. Perhaps the best way to explain the contest details is to simply quote from the Official Contest Rules:

RULES AND GUIDELINES

1) Teams must consist of two or more students. Middle and high school teams compete separately. Students should teach at their own grade level or lower.

2) Each team will develop a one-period class lesson focusing on a science and/or health topic.

3) Entry forms will be judged on the basis of originality, use of hands-on experiments or role playing activities that involve many students, and how well the lesson teaches important concepts.

4) Each teacher may encourage as many teams as desired, but can submit a maximum of four entries.

5) After reviewing the entry forms, SEP will select approximately 20 middle school and approximately 20 high school entries as finalists. Each finalist team will present its lesson in a real classroom and will be observed by a team of UCSF judges. Lessons may be revised up to the time they are taught.

6) Prize winners will be announced at an awards ceremony at UCSF.

The contest begins in early February with publicity focused to attract students and teachers, and it ends with an awards ceremony in the third week of May. We usually allow about six weeks for students to learn about the contest, form teams, and develop and submit their entries. We notify the finalists before spring break, and then the teams have at least three weeks before the judging period. Each team makes one official presentation during the three-week judging period. Students make their own classroom arrangements, while SEP staff coordinate volunteers' schedules so that three judges attend each presentation. Winners are announced at the awards ceremony, and the level of tension and enthusiasm escalate to a level normally reserved for athletic events. The contest process requires a great deal of staff coordination. After a brief recovery period, we remember the enthusiasm of the student presenters and lesson recipients, and we conclude that the effort is a wise investment.

Quite frankly, the prize money is the initial motivation for many students. We award $5,000 as follows:

> FIRST PLACE: $1,000 ($500 for the team; $250 for the sponsoring teacher; $250 for the science department; one middle school first prize and one high school first prize)

> SECOND PLACE: $500 ($250 for the team; $125 for the sponsoring teacher; $125 for the science department; two middle school and two high school second prizes)

> THIRD PLACE: $100 ($50 for the team; $25 for the sponsoring teacher; $25 for the science department; five middle school and five high school third prizes)

In keeping with the American tradition that no event or organization exists unless it is emblazoned on a piece of wearing apparel, we print and distribute a SEP Contest T-shirt to every student who enters. This

T-shirt features the names of all the participating students, teachers and schools. At the ceremony, we also award a prize (science book or, better yet, hands-on kit) to a Best Science Student nominated by every district middle and high school. After the ceremony, we serve refreshments at a reception that is lavish by public school standards. Contest supplies cost approximately $11,000. We have found that it is relatively easy to raise money for contest supplies and prizes. However, we have not been able to raise money specifically for the staff time involved, but the contest does help generate enthusiasm and support for the program from the entire community, including potential donors.

The most frequent criticism we hear is that the contest, by its very nature, excessively rewards competition, and that the "losers" finish with some negative feelings. Given the constraints of the format, we do everything we can to overcome this drawback. All the students receive a T-shirt that has their name listed equally with everyone else. All the finalists are publicly recognized at the awards ceremony which also features an enjoyable keynote presentation by a local celebrity such as a television meteorologist, a Nobel laureate, or a resident Mr. Wizard. We have also discovered that, by the time the contest is over, most of the student teams have enjoyed researching and preparing their presentation, are thrilled that they survived the process, and are much less attached to winning the money.

The initial entry form solicits the necessary student and teacher addresses and phone numbers, and asks students to briefly describe their proposal. They need to tell us what they want the "recipient" students to know or be able to do after the lesson; what they will do as teachers in presenting the information; and what the class will do (activities) during the lesson. Emphasizing the activity focus of the lesson, we specifically ask the students to provide details on experiments or role playing or "games." Submissions that are not selected as finalists usually omit this aspect or simply say, "We are going to do an activity with acid rain." Unfortunately, students whose teachers emphasize vocabulary usually make presentations that are dry and filled with disconnected facts. The best presentations involve recipient classes in experiments and lively discussions. In some ways, students, as teachers, have more leeway than adults. They involve the classes in singing rap songs, wearing costumes, and talking about personal issues.

We try to help student teams by matching them with university health professionals and scientists. Teachers who have university partners often use them in this activity. It requires comparatively little input to make a big difference in the students' presentations, especially by critiquing their first drafts and their practice presentations. The most important role that university volunteers play is in judging the presentations. We ask each volunteer to judge at least three lessons, and we try to arrange to have three judges at each one. The winners are chosen at a judges meeting where we all compare notes and discuss our observations. A staff person, who has seen as many presentations as possible, coordinates the meeting. We do not pretend to have an objective system that exactly grades the lessons. There is simply too much variety in topics and grade levels of presenters and recipients. Nonetheless, we have always felt comfortable with our eventual decisions.

From a programmatic point of view, the contest helps enlist new volunteers, creates enthusiasm among all constituents, and provides an exciting climax to the academic year. It also provides an enjoyable mechanism for program staff to match faces and bodies with all those voices on the phone. We have considered using the contest to begin the academic year rather than conclude it. San Francisco middle school teachers, in particular, seem to get swamped with student activities (Science Fairs, Science Olympiad, Invent America) in the spring semester. The drawback to a fall contest is that many students need the first semester to upgrade their science skills before attempting to research and present their own lesson.

Because so many health issues have a strong scientific component and because we are a health sciences campus, many presentations have focused on health topics such as sex and drugs. Since it is more difficult for students to perform acceptable class experiments on these topics, the activity aspect often focuses on role-playing simulations or demonstrating birth control techniques. In one lesson, student presenters distributed condoms to eighth-graders, and the class then put the condoms on anatomically detailed models. The sensitive nature of these topics may cause some concern on the part of teachers, parents, students, principals, university staff and the community. Despite such potentially controversial presentations, we have not had any complaints. If you live in an area that is less permissive than San Francisco, you may need to exercise more control over the content of some of the presentations.

The Lesson Plan Contest is one of our most popular and satisfying activities. It genuinely increases student interest in science. Students invariably remark that it gives them a greater appreciation for their teachers and for science teaching as a profession. "You mean she does this every day?" is a typical comment. From a program point of view, it is one of the easiest ways to enlist new volunteers and get them excited about SEP. Watching the students present lessons or cheer for their colleagues at the awards ceremony is also one of the best rewards that we experience.

SCIENCE AND SOCIETAL ISSUES SYMPOSIUM

BY SUSAN BRADY

Students from Analy High School in Sebastopol are presenting a soap opera skit to an audience of scientists, educators, consumer advocates and students. A young married couple and a 35-year-old single mother are deciding whether to have their babies tested for cystic fibrosis and Down's syndrome. In the next room, students from Clayton Valley High School are dramatizing a civil war that has erupted between Northern and Southern California over water use. Across the hall, students from Walnut Creek's Northgate High School are enacting a courtroom drama in which Holland Pharmaceuticals is defending its use of laboratory animals. In another meeting room, students from Ukiah High School pretend that they are members of a United Nations committee analyzing the use of genetically engineered agricultural products to prevent famine.

These scenes took place at the Lawrence Hall of Science in Berkeley on Dec. 9, 1989. Since 1986, LHS has sponsored an annual Symposium on Science and Societal Issues for high school students. Each year, teams of students from 30 schools throughout California meet to present recommendations about societal and ethical issues related to advances in science and technology. Rather than deliver abstract lectures on their chosen topic, most teams design innovative formats so they can present the information in a context that is embedded in a real world situation.

Each September, LHS mails a list of topics (see description of samples chosen from among the 10 topics in the 1992 contest) to schools, corporations, universities, government officials, and environmentalists. Scientists who volunteer in schools can help students prepare for the Symposium. For example, scientists who participate in the STEP pro-

SCIENCE EDUCATION PARTNERSHIPS

gram (see the article "One-on-One Partnerships" on p. 26) are encouraged to work with their teacher partner's students. LHS staff also provide the teams relevant articles, reference lists, and names of individuals from universities, private companies, regulatory agencies and consumer groups who have agreed to be available for personal interviews.

During the day-long Symposium, each student team makes its presentation twice, to two sets of judges. In the afternoon, the top three teams from the morning competition repeat their presentation for the entire group. All teams receive prize ribbons, and each student receives a certificate of participation. Scientists from industry and universities, representatives of government agencies, elected officials, high school students and members of environmental groups act as judges. The judges are available during lunch to interact informally with students and discuss topics and careers in science.

In its six years, the Symposium has involved more than 120 secondary science teachers and over 1,500 students from nearly 100 high schools throughout California. More than 350 university and industry scientists, public officials and environmentalists have participated as resource contacts and judges for student presentations. Many Central and Southern California schools expressed an interest in the Symposium, and some even traveled more than 400 miles to participate. As a result of this interest, a successful Symposium was held in April 1992 in Los Angeles in addition to the LHS event. A volume of Symposium Proceedings, which includes a two-page summary of each presentation as well as team photos and participant lists, is compiled, published and disseminated to all participants.

Involving high school students directly with ethical and societal issues that arise with new scientific advances, gives them a chance to develop important problem-solving skills. To prepare their presentations, students conduct library research and interview experts. As students work as a team to formulate a group presentation and topic recommendation, they develop and practice teamwork and critical thinking skills. Students learn that science profoundly impacts their lives and that learning about science can be fun. Scientists who serve as judges and resource contacts consistently report they are impressed with the quality of the presentations, and the poise and intelligence of the students.

SELECTED 1992 SYMPOSIUM TOPICS

Fetal Tissue Research

Recent research has shown that cells obtained from aborted human fetuses may be used to cure diseases. What limits, if any, should be placed on the use of fetal tissue for research or medical treatments?

The Spotted Owl

Heated controversy has surrounded the issue of habitat preservation for the endangered Northern spotted owl in old-growth forests. What are these habitat requirements? What compromises can be made among loggers, environmentalists and economists? Who should make decisions about the balance between ecological and economic interests?

AIDS

Many AIDS sufferers are eager to try new drugs even though formal testing of such treatments are not completed. Should terminally ill patients be permitted to participate in human experimentation with treatments that are not approved by the FDA? How should such research be regulated?

Use of Animals in Research

What types of research use animals? What are the alternatives to animal use, and how effective are they? Who should regulate research procedures that use animals?

EXCERPT FROM A VALLEJO SENIOR HIGH SCHOOL PRESENTATION

Mark Twain once said that "in California, whiskey is for drinkin' and water is for fightin' over." Twain's observation has proven to be startlingly accurate in the twentieth century as inhabitants of the Golden State fight tooth and nail over this aqueous treasure.

Governor Eb E. Nezer is an uncaring, uncompassionate clod . . . a real scrooge, if you will. As Governor of California, his only concerns are getting re-elected and what's on the menu at the annual Christmas banquet. Presently, he is viewing a fictitious bill which pretends to solve California's water problems but falls quite short of its mark. The governor is visited by a representative of the agricultural, urban and environmental sectors who each wish to make him aware of their disagreements with the bill. After Governor Nezer rudely removes the three individuals from his office, he drifts off to sleep and is "visited" by the three representatives.

The environmental "apparition" is particularly effective in describing the scientific facts associated with California's water situation. In his sleep, the governor learns about the water cycle, aquifers, bays, and wetlands. When he awakens, he suddenly realizes that the inadequate water legislation must be vetoed and that the answer to California's water woes can be found in extensive water conservation. He proposes a strict regime that uses water wisely. The governor's awakening is symbolic of the fact that all Californians must wake up and open their eyes to the truth surrounding California's water, at which time we will be one step closer to a solution.

SUMMER RESEARCH OPPORTUNITIES FOR STUDENTS AND TEACHERS

BY MARGARET CLARK

For the past few years, SEP has coordinated a summer internship program that provides the opportunity for high school students to directly experience scientific research. The program started small, and grew as funds and laboratory slots became available; this past summer, 16 students and two teachers were placed with UCSF investigators. This is mostly a program for the more academically motivated students who may be interested in pursuing a scientific career and primarily targets minority and economically disadvantaged students. We do not necessarily select the most academically outstanding students; maturity can be more important than academic performance. A priority of our program is to provide an opportunity that could make a real difference in students' futures. SEP tends to give priority to sophomores and juniors, because by participating in the program students may be encouraged to take additional science courses and they will serve as inspirational role models while still in high school. However, we often support seniors, especially if a lab wants a particular student back for a second year. This gives the already trained student an opportunity to contribute more significantly to the lab research program and feel a strong sense of accomplishment.

The SEP program at UCSF is funded from several sources. The National Institutes of Health Minority High School Student Research

Apprenticeship Program is our largest single source of support for students. We also receive student stipends through American Chemical Society's Project SEED, and through the Lange Fund of the UCSF Department of Physiology. Additional funding has been made available by several individual investigators who provide full or partial support for SEP summer students working in their labs, thereby giving more students this opportunity.

Although nominally a summer program, the summer internship program actually requires year-round effort to implement. Putting funds together for the coming summer starts in the fall, and we are confirming laboratory slots in early spring. We solicit nominations of students in early spring by sending a description of the program and criteria for selection to teachers who have participated in SEP programs. We select students for initial screening interviews by SEP based upon a letter of recommendation from a sponsoring teacher, together with a resumé and letter from each student. SEP typically interviews twice as many students as the number of available slots.

Based on the screening interviews, the SEP coordinator selects a subset of students who appear to best fit the program priorities, referring two or three students to each laboratory that has offered a position. Making lab referrals is a little like match-making. The coordinator tries to see that the labs are happy with the students sent to them, while attempting to place all the students who could most benefit from the program. Lab mentors and sometimes the principal investigators interview the students referred to them and notify SEP of their order of preference and whether they have reservations about taking any of the students. Students who are selected by more than one investigator are given their choice of projects. This part of the process takes time because the investigators are busy and may not schedule interviews right away. Occasional phone calls will probably be necessary to move the process along.

Students in the SEP program work 20 hours a week for eight or nine weeks, depending on the amount of the stipend provided by their funding source. They work half-time because supervising a high school student requires considerable time and effort. We want to make sure that lab mentors have enough time to focus on their own work without distraction or obligation to keep a student busy. It is also important that the students gain actual research experience. We emphasize to the labs that the student interns are not there simply to wash glassware or prepare solutions, but to engage in work that involves responsibility and development of lab skills. Because of inexperience, students start slowly, though we have had students close to publishing results from their work by the end of the summer. We also have found that they perform best when treated like adults and are expected to be responsible members of the lab team.

To provide a perspective on their individual research experiences, SEP schedules weekly meetings for all the students. A major benefit of these meetings is that the students get to know each other and share experiences of getting started on their projects. It is also an opportunity to present a general orientation to the university and facilities, such as the library, to handle business, and to get feedback from the students about how things are going. In addition, invited guests talk about topics of special interest, or students can take special tours. For example, last summer's program focused on molecular biology. Students had discussions with scientists about sickle cell disease, genetic engineering and about the Human Genome Project, followed by a two-hour, hands-on tour of the UCSF Human Genome Center. At the end of the summer, weekly meetings provide an opportunity for the students to give brief presentations on their work. They write a short summary describing what they have done and what they have learned or gained.

SEP also provides laboratory internship experiences for a few teachers. Because of their greater experience, the teachers work more independently than the students. A summer in the lab provides teachers an opportunity to learn firsthand about current research interests. Many science teachers have never engaged in scientific research and their only laboratory experience has been in the artificial setting of college courses. The summer internship is a powerful means of supporting teachers as science professionals and rejuvenating their enthusiasm for science. In addition, it can lay the foundation for a partnership between the laboratories and the teachers' classes during the subsequent school year. The teachers may take their students on tours of the lab, and members of the lab visit the classroom. In this way, sponsoring one teacher in a summer research position benefits a large number of students.

Overall, a summer research program — though benefiting only the limited number of students who participate — is a valuable component of a school outreach program. The encouragement that students receive through interacting with members of the lab group can be a key factor in their decision to pursue a scientific career. And the professional scientists are often equally inspired by the bright, eager students they mentor, perhaps because they provide an opportunity to pass on the kind of support the scientists received at a similar stage in their lives.

MY SUMMER IN THE ALBERTS LAB

BY MOSES KIM

The summer after my junior year in high school, I worked in the Alberts lab at UCSF. The lab is studying a filamentous protein in Drosophila (fruit fly) embryos. This protein, called intermediate filamentous (IF) protein, is present in all cells of most organisms. The protein forms a lacy network throughout the volume of the cell. It was thought that the filament supported the structure of the cell, but a test proved this wrong. Since the function of the protein is unknown, the lab is trying to find out whether the protein has a function in the early development of an embryo.

Bill Sullivan, who studies this protein, taught me more than anyone else in the lab. From darkrooms to flyrooms, he and the other people in the lab taught me many techniques that are used in a Drosophila lab. I not only learned how, I learned why.

Let me briefly explain one experiment I did last summer and the techniques I used. We studied and experimented on an antibody to determine if it binds to the IF proteins of Drosophila. We used two techniques, immunofluorescence and western blotting.

Immunofluorescence is a way of determining the location of a protein in a cell or a whole organism. You use an antibody that binds specifically to the protein that you are studying. The first step in immunofluorescence is to fix a sample (I was fixing Drosophila egg chambers) and incubate it with the antibody of interest. Then incubate the egg chamber with a different (secondary) antibody that specifically binds to the primary antibody. This secondary antibody contains a special molecule that glows under UV light. That way you can see where in the cell the primary antibody has attached itself to the protein that you are studying.

To be able to stain the egg chambers with antibody, you first must isolate the egg chambers. Because the egg chamber is the younger stage of an embryo, the egg chamber must be forcefully removed from the female fly. To accomplish this, you first must make the buffer and the fix (luckily, I never had to make these myself). The next step is to gather all the needed materials: three beakers of decreasing size, a blender, ether, a mesh with grating smaller than an adult fly, and 2-4 bottles of flies. The flies are etherized and then chopped up in the blender. Then you use the mesh and the beakers to separate the egg chambers to study them further.

Then you fix and stain the egg chambers with the antibodies. You look under the microscope and photograph the best samples. After the egg chambers have been photographed, the film is developed and printed in the lab.

By using the techniques that I learned this summer, one is able to see the distribution of the protein in an embryo or an egg chamber and also determine the size of the protein. Since the protein can now be seen, the "average" scientist may hypothesize on the function of the protein and conduct further experiments.

There are many new things I discovered this summer, but I would like to tell you about some of the discoveries I will never forget. I discovered that science isn't all technical; it is also fun and exciting. I also discovered that scientists aren't all sober-faced men driven by their thirst for knowledge; they are human and aren't much different from me. Most are very nice and have great patience. They also have a lot of fun in and outside the lab. Sometimes they are bored and frustrated because that's how science is, but they take it very well. Another thing I will never forget is how a lab is run. Many people think that in a lab, everything runs easily and smoothly. This stereotype is false because not everything is easy. There are many hard techniques you have to learn. Also, not everything goes the way you want it to.

Even though science isn't all fun and excitement, I enjoyed my summer and I will not forget the people I worked with for two months.

Del Mar science teacher, Larry Flanner, analyzes data for space shuttle biology experiments in his IISME-sponsored summer employment at Lockheed (photo courtesy of IISME).

(Right) Sylvie Plamondon, a high school chemistry teacher, updates her laboratory skills and knowledge base in a summer program (photo courtesy of IISME).

EVALUATION & RESOURCES

PERSPECTIVE OF A PROGRAM EVALUATOR

THE TURING TEST AS AN EVALUATION METHOD

RAISING MONEY FROM THE PRIVATE SECTOR

THE GRANT DONOR/RECIPIENT PARTNERSHIP

RAISING MONEY FROM THE PUBLIC SECTOR

RESOURCES FOR SCIENCE EDUCATION PARTNERSHIPS

PERSPECTIVE OF A PROGRAM EVALUATOR

INTERVIEW WITH MARK ST. JOHN

Art Sussman: What are the common attitudes that you encounter about evaluation?

Mark St. John: The word evaluation has some unfortunate connotations but in all these years I haven't come up with an acceptable replacement. Evaluation is usually regarded as aiming to prove things. Typically, it aims to measure in a kind of undisputable way that you have made a big difference, that the funder has made a great investment. Perhaps it essentially goes back to our notions of being in school and being graded, proving that you are a good person because you received this good report card.

Now the person who gives you the grade has to be objective. As a result, the evaluation is typically regarded as being independent from the project. A summative evaluation would objectively measure the state of affairs before and after the project and compare these measures with a control group that did not participate in the project but was otherwise identical to the project target audience. The evaluator in this model should not have any connection with the project staff. I think this model is an imitation of what people think physical scientists do. Actually, physical scientists do not operate that way very much, they do a lot of messy stuff. But these summative model evaluations aim to imitate this rigorous model of physical science. Unfortunately, this model is not so appropriate or useful in the complex interactions in which our grant projects are immersed.

AS: Do you think that people who are applying for grants are coming from that perspective because they believe in that kind of evaluation, or is it because that is what they think the funders require?

MSJ: Generally, they think that is what the funders require. There is in this business a kind of paranoia, on both funder side and practitioner side, of thinking that there is someone out there who is demanding hard, quantitative rigorous measures. Yet when you talk to any of these individuals, they almost always blame it on the other side. I talk with funders and they are saying, "For some reasons, these guys want to make all these measurements." I talk with the grantseekers and they say, "The funders require this and my board needs this." My sense is that people overestimate the need for this kind of data, as opposed to a bolstered argument. If one wants to understand what you are doing and its impacts and its operations and you want that argument grounded with evidence, that is very different than this kind of pseudorigorous proof.

There is in this business a kind of paranoia, on both funder side and practitioner side, of thinking that there is someone out there who is demanding hard, quantitative rigorous measures. Yet when you talk to any of these individuals, they almost always blame it on the other side.

This notion of evaluation also is connected to the way that grant proposals are structured. Many excellent projects start out much more experimentally than the way we write them in the grants proposals. Often the key is simply getting very talented and committed people together on a daily basis, having some broad objectives and giving them the resources to go in and do everything that they can. In contrast, a grant proposal has a formalized description of the project with definite goals, and specific, measurable outcomes. This story version of a project tries to prove that we have made this measurable, significant difference, and therefore the funder's money has been well invested.

At the bottom of it, everybody has in their heart that they want to make a difference in the lives of kids and the lives of teachers. A good project should make a noticeable difference. I think some of the frustration and paranoia associated with evaluation results from the fact that many of the efforts that we make do not seem to have changed what is happen-

---

ing in the schools.

AS: What kinds of things do you think would be realistic to write into a proposal?

MSJ: If I were a funder today, I would want to see someone who could convince me that the core idea that they have is inherently a good, powerful idea. Today I think people tend to think that there are certain forms that you go through and if you go through all of those forms, it will come out good. That is not true, you have to start with a good idea, then you have to shape it. So anyway, that is the first one. Then I would want to see that they are responsible versus accountable. Now if I thought they are trying to be accountable, that would tell me that they are trying to meet what they think my criteria are and they are trying to satisfy or prove that what they are doing is good, but they are trying to prove it in my terms, not their terms. I don't know what their terms are.

So I would rather see them be responsible, saying we are thinking that we are going to try this, and we are thinking that in all likelihood, given what we know about the world and what we can do, these kinds of things will happen. Our key concerns are these three specific questions. For example, Do the teachers have enough content to really become inquiry learners? Or do teachers have enough freedom in their classroom when they get back to carry out some of these things? If not, what kind of support can we give them? We are going to launch our initiative and form a teacher research group and investigate these three key questions during the first year and we are going to keep pressing on this to maximize the degree to which we can influence what happens in classrooms. Therefore, our research is aimed at what we think is the hardest question we are facing, which is to make real change in the classroom. So there is a learning kind of research, it is very real, it tells me what they think are strengths, what are weaknesses, very realistic expectations and it says if you invest in us, we are going to do the best we can and we are going to continue to learn about what we think are the real hard issues and we will share what we learn.

AS: Then in the proposal they say they are going to use Mark St. John as an evaluator. What will Mark St. John do?

MSJ: We are going to call in an external evaluator to help us in this process of learning. Basically, here is the notion. We see our project not as a sure thing but as an experiment. We think we have good people, we are going to do the best we can and we are going to use Mark St. John to document what happens here so that the experiment can be shared with others and we can get as much insight into what is happening. He is an insightful guy and is going to bring in a perspective that provides a bigger framework, to be able to see bigger patterns in this and relate it to bigger issues. But he is not just going to document. He is also going to facilitate our own workers and our own teachers in internal processes that are going to help us learn along the way and get smarter and change our plans as we go. So he is going to facilitate a learning process that rather than us knowing all of the answers ahead of time, we have a community that is learning and the end outcome of this whole project may be a community that is slightly smarter and more capable than it was before, so as we continue in our projects, we are going to be a whole step up from where we were before.

I look at the evaluator not in terms of an isolated project, but rather in terms of the communities that are involved. Typically, the evaluator is seen as someone who comes from outside the community, gathers certain kinds of information through a survey or tests or interviews, and then leaves the community. He writes a report for an external audience saying this is what has happened in that community and this report is primarily for external audiences.

I see a different role where the evaluator is part of the community, is in the midst of it, yet is distinguishable, has a clear role, and has a clear separation. He does three functions: he imports information into the community, he facilitates exchange within the community, and he exports information to the outside community.

Consider a hypothetical program that brings together a variety of institutions to increase the interest of disadvantaged youths in scientific careers. The evaluator would bring information into that community based on his knowledge of research reports, other model programs, the broader picture of educational reform, etc. Second, he would facilitate the exchange among the institutions that are involved. He would provide an additional network node and would facilitate the learning that occurs as the project progresses. Finally, the evaluator will produce reports and briefings for a wide range of audiences, including the funder.

AS: As you move away from the role of the evaluator as an external auditor to this role of facilitating, how do you deal with questions of bias? How can you produce documentation that would satisfy an external audience?

MSJ: Exactly. With this different model, people might wonder — this guy is going to be co-opted, he is not going to be impartial, how are we ever going to find out if this project is any good? These are legitimate concerns of the funder. The evaluator has to retain an independent perspective and provide honest feedback. In the documenter aspect of the role, the evaluator functions as an educated connoisseur and a critic. A good critic is not necessarily critical in the negative sense, but portrays what is going on in ways that allow you to get new insight. Essentially, you have to trust the expertise of the evaluator to facilitate and still maintain an independent focus.

Let's take this a little further. I have a lot of experience looking at schools and projects, so I can go in and see things rather quickly. I suppose like a chess player looks at a board and immediately sees patterns, I see educational patterns rather quickly. People say how did you know that, and I say it's because I know the pattern, and this situation is not unique.

> *Being an evaluator is a wonderful way to develop people's ability to see situations in different ways. So if you take teachers and give them that role for a while, then they have a whole different perception of education.*

Well, there is a connoisseurship that develops and I think that the flip side is that being an evaluator is a wonderful way to develop people's ability to see situations in different ways. So if you take teachers and give them that role for a while, then they have a whole different perception of education. So you can treat evaluation not only as a way of portrayal, but also as a way of learning to become much more skillful viewers of and participants in the educational scene. Again evaluation becomes a vehicle for everybody getting smarter, and that is another way to break down the old model of the independent, solo evaluator writing his external report.

Returning to the funder's concerns, an interesting feature of this model is that the funder is also part of the community. The funder has particular questions that they might want to answer. Now my sense is, if I give the benefit of the doubt, most funders really just want to be a little bit smarter about investing their money. What they want is to be privy to ways to invest money that have greater positive impacts. How can we maximize the use of the money? The evaluator can take this question, help clarify the specific objectives of the funder and very specifically tailor things to those questions. Funders sometimes go crazy and think they need rigorous proof. But they just have real questions, just like people in the system have real questions and if you press them a little bit, the real question is we want to do something about the education system, how can we do it? What particular role can we take? So here you have interaction between the evaluator and the funder and the community which says let's try to find out the answers to your questions the best we can. The funder, the community and the evaluator work together to help create the substantial change that everyone wants to see.

AS: Do you administer any quantitative pre- and post-tests; when and why?

MSJ: I do it from time to time, especially if I have a specific question. My goal would be to get insight into the situation, rather than think that I had rigorously proved something. The trouble with proving something is suppose that I prove that this curriculum works very well in this situation, now what does that buy me? It buys me almost nothing. I know nothing about its use in other situations and if someone looks very hard, they could probably pick apart my proof because it is extremely difficult to set up rigorously controlled experiments in schools. How do you isolate one variable and how can you have an exactly matched control population?

AS: Yet there is a national effort to identify curricula that have been proved to work and then disseminate those lessons.

MSJ: I have mixed feelings. I have doubts about this over-rationalized model of let's test the curriculum and prove that it is effective, and once it is proven effective like a medical drug, we can disseminate it around and it will work for everybody. My opinion is that the context is 95% of the game and the thing that you put in is 5% of the game, so any curriculum that you insert is immediately assimilated into the existing system.

Even the notion of implementation comes from the same model — develop the product, do the research, test it and then disseminate it. But that is just not the way that it happens. I think it is fine to test a new curriculum and learn something about its use, and see how it can be applied in other situations. But this notion of proving it beyond a doubt, I just don't find that very interesting.

AS: In subatomic physics we learned that we cannot separate the observer from the experiment...

MSJ: Nor can you separate the tool from the carpenter. It is like certifying a hammer and saying this hammer is going to work perfectly in all situations. But there are lot of different kinds of carpenters around, and yes there are better hammers and worse hammers and you can certify them, but it is only 5% of the variance. The carpenter is 95% of the variance. That is what we are talking about. Actually, it is trickier than that because you can do professional development for the carpenter, but then you realize that he has a whole set of traditions that he is operating within. So I keep thinking that there are much deeper constraints, what I call scripts and traditions that determine things, much more than we are getting at. We keep getting at the surface layer of education, like training and skills, like assessment measures and curriculum, but fundamentally those do not determine what happens in schools.

AS: I would argue that you don't do anything in isolation, that part of changing science education is connected to radically changing schools, which is connected to fundamentally changing society.

MSJ: Absolutely. And I think there is a ceiling of what you can do in science education without making radical changes, and that is what people are discovering, now that they are talking about truly restructuring and making systemic changes. The culture of the schools has become very, very fixed. Just like when you go to a ballpark, there is a way that you get hot dogs, and you stretch in the 7th inning; schools are the same way. There is such a fixed way of being, it doesn't matter what you put in there, that way of being predominates and that is what I find so amazing. At first you might blame the teachers, but you notice no matter who you put in there, they act the same. So after a while, you begin to think that there is something that is determining behavior beyond individual attitudes and skills of that person. If you give someone hundreds of hours of workshops, they are not noticeably distinguishable. They don't stand out, they are not doing something completely different from everybody else. In fact their classrooms look very much the same, maybe a little richer, a little smarter, but incrementally, not fundamentally. So then you begin to think something deeper is determining what is going on.

An analogy might be that there is a play happening and a script is written and the stage is set and now we are trying to get professional development to the actors. Well, they may get a little better in their roles, but their roles are the same, the words are the same, and the set is the same. So how much can professional development accomplish without saying maybe we should throw away the script, maybe we ought to get rid of the set, maybe we ought to make it up as we go along. I don't think we see very well what that script is, what the stage set is. It is such a given, the water we are swimming in. Coming back to evaluation, that is where I see evaluation playing a role in a much broader sense, saying let's discover together the water we are swimming in. Let's begin to think about and revise our fundamental assumptions.

THE TURING TEST AS AN EVALUATION METHOD

BY MARK ST. JOHN

One of the interesting and controversial areas in the field of artificial intelligence revolves around the question of how you would know when you have developed a machine that can think like a human being. In 1948, Alan Turing suggested a simple way to decide whether machines are capable of thinking like humans. His Turing Test involved a kind of thought experiment in which there were two rooms — a computer inside one, a human in the other. Questioners could submit any written question through a slot in the doors and, after a while, a written answer would come out. Questioners could ask whatever they wanted, as long as it could be put in writing. If the questioner could not reliably distinguish which answer was the human's and which was the machine's, then, according to the Turing Test, machines could "think." The key ingredients to the test include the following ideas:

The questioner makes blind comparisons.

The questioner is free to use his or her skills, expertise and connoisseurship to help devise ways to heighten the differences and make discriminations (e.g., the questioner could ask the machine and the human to make analogies, recognize a drawing, or carry out complex calculations — all of which are likely ways to help make the distinction clearer).

The criteria by which the distinctions are to be made are not specified in advance — expert judgment and exploration are used to distinguish between the compared cases. The distinctions that are noted are reached in a goal-free, rather than a predetermined, way.

It occurred to me that the Turing Test provides a simplified basis for designing ways to assess systemic change. The test allows for a multifaceted, sensitive, responsive way to examine the extent and the ways in which a funder's investment is indeed fostering changes in all levels of the system — from institutions, to schools, to classrooms, to individual student experiences. In the artificial intelligence application of the Turing Test, one studies the output to distinguish between a human and a computer. In this education application, one uses the same technique to ascertain whether our educational investments are making a distinguishable difference in the classroom and in students' experience.

In brief, we send expert investigators into classrooms. The investigators do not know the background of the classroom situation; e.g., whether the teacher has had extensive in-service training or not, or whatever the variable is. The question is: Can the investigator distinguish differences in the classroom experience. Two examples would be:

> Could skilled educators distinguish the classrooms of teachers who had been through extensive professional development programs?

> Could classrooms in states that have frameworks and statewide assessment programs be distinguished from classrooms in states that are nonregulatory?

It is also important to point out that the Turing Test not only can answer the question as to whether the two are distinguishable, it also can provide illumination about the ways in which they are distinguishable. By fostering an atmosphere of exploration and discovery, this technique allows truly important differences to emerge.

This test is fundamentally different from a goal-based measurement and comparison system, and from a purely qualitative examination of program impacts. I investigated this approach when a foundation executive asked me to evaluate the foundation's summer teacher institutes, to do a study to prove that these teachers have obtained new skills, attitudes and ways of teaching that are making a difference to the students' learning experiences.

We decided to see if we could do something basically in the middle. By the middle, I mean on the one side you can just ask the teachers if they had a good time at the institute and they say yes. On the other side, you can measure the SAT scores of students to determine if those teachers are much better than the others, but that is unrealistic.

We decided to compare master teachers who have had hundreds of hours of training, those who attended a couple of summer institutes and those who have had no professional development. Looking at those teachers, if professional development is a significant influence on teachers, we ought to be able to find some distinction between them. I don't know what the distinction is, but if they are completely indistinguishable in every way, then we would say that the professional development doesn't seem to make much of a difference.

So the rules are that you can ask any question that you want, or make any measures that you want with this set of teachers, except ask about their professional development background, because that is the blind variable. As a connoisseur, you should be able to figure out what are the things that might be noticeable. For instance, maybe a crucial difference involves students working in groups, so I am going to document to what extent students work in groups. Or maybe they are teaching advanced topics, so I will give a test on those advanced topics and see which students do better.

You can do anything that you want with these teachers, except ask them about their background. In the study that we did, we videotaped their classrooms, and had them keep logs of the kinds of things that they did — whether they used groups or not, how many demonstrations and labs they did. We talked to them about their philosophy of teaching; we pre- and post-tested their students on content. We asked them to keep copies of student portfolios, and to select the most creative work that students did, all those kinds of things.

In this particular case, the context was high school chemistry, and at one point I was visiting five teachers in two different states, luckily in the same week. When I went to those five classes and videotaped and talked to them, the classes were nearly indistinguishable. They were teaching the same topic in the same way on the same week. This is what I call the unwritten script, they are all doing the same thing in very much the same way. They are at the blackboard talking, and they are asking simple questions, kind of recitation things. So I made the videotapes and shared them with experts, asking them to evaluate these teachers. And the evaluations did not correlate at all with staff development.

Another interesting thing is that I interviewed groups of students from each class. So I took four students and I would sit down and ask them about their experience of learning chemistry: What is chemistry like?

Do you follow the book closely? When a student has a question, do you go away from the book? Are you on a forced march? Is chemistry problem-solving or is it understanding the world? Do you study issues in society? Do you work as groups? I asked everything I could about their experience of learning chemistry. Once again, if I showed you the videotapes of those kids talking, you would find them indistinguishable from each other. The biggest variable was whether they liked the teacher, whether the teacher was a kind person to them. Basically, that overrode the negativity of learning chemistry. The good teacher helped them through this horrible experience.

So through all of that, you begin to wonder what are these hours and hours of professional development doing. I think where you can begin to see it show up, is basically they are helping the better teachers infuse occasional diversions into the standard script. So as we do this covered march, as we go through the play and follow the script, I can infuse every now and then a little bit of something or perhaps when a student asks a question, I have two ways of answering it, so I am a little bit richer. But in terms of systemically transforming the experience of high school chemistry, this is not happening as a result of summer institutes and professional development for teachers.

One point to note is that we were careful to pick for our control group teachers who were known as good teachers, but who were not engaged in professional development. At its simplest level, the study indicated that the professional development did not significantly change the classroom experiences provided by high school chemistry teachers who have a reputation for being good teachers. This does not necessarily mean that staff development is a waste of time and money. Maybe some teachers are good chemistry teachers because they take staff development and that keeps them motivated. Another subset of good chemistry teachers uses other techniques to be or remain good educators. Maybe we need staff development plus other factors (10 times the amount of money for supplies and equipment, a radically revised curriculum, a restructured school day, dramatically smaller class sizes). Nevertheless, the result challenges our views about how to make systemic changes in schools. Further, this Turing Test provides a method to evaluate educational interventions especially with regard to systemic change.

RAISING MONEY FROM THE PRIVATE SECTOR

BY DENNIS HARTZELL

The most compelling vision and the most altruistic intention will, sooner or later, come face-to-face with financial realities. This encounter is frequently wrenching for all who share the vision and celebrate the intentions. More often than not, the financial realities are harsh and unforgiving.

The Science and Health Education Partnership (SEP) at the University of California, San Francisco began as an initiative driven entirely by the time and energy, freely given, of several senior faculty members. The very small startup costs were easily handled by their access to modest discretionary funds. Enthusiastic volunteers worked directly with teachers and students. Recognizing SEP's exceptional and potential contributions to the community, UCSF's chancellor provided a relatively small but nonetheless vital sum to help seed the program.

SEP's immediate success in enhancing the experience of public school science teachers and their students made it clear that an infusion of funds could dramatically increase the program's impact. Volunteer programs serviced by an administrative assistant "on loan part-time" from other duties are limited in what they can accomplish. SEP's potential could only be realized by a level of funding beyond the reach of the campus' own resources, especially in a time of fiscal constraints. The money to fuel SEP's growth would have to come from external sources.

The process by which we set out to raise money for SEP is a model of fundraising at its simplest and most direct. Our approach consists of four steps: identify potential funders, develop materials, submit letters of inquiry and grant proposals, and maintain good relations with donors.

IDENTIFY POTENTIAL FUNDERS

We compiled a list of prospects. We looked for foundations and corporations that identify K-12 education, especially science-related, as a priority for their grant making. We secured as much information on these prospects as we could, using annual reports, the Taft directories for corporate and foundation giving, and the *Guide to California Foundations*.

Geographic considerations are important in choosing a prospective funder. We examined local foundations. We also looked at local major corporations, especially those that have a scientific or technological base.

One problem that you are likely to encounter is competing priorities at the institution. The majority of potential local donors that we identified were already targeted by other programs at the university. Since many of these programs carried a high priority at the university or had long-standing relationships, we had to eliminate these names from our list. You do not want to make enemies at your institution, and you do not want to confuse the funders. For example, SEP's staffers wanted to approach a "reserved" major donor for a relatively small amount of money ($15,000), feeling that it would help SEP a lot but would not really compete with the university's major approach to the funder for $250,000. This could not be allowed. If the funder is wavering, you do not want to give him a cheap excuse to say no to a major request.

DEVELOP SUBMISSION MATERIALS

We developed materials that could be used in letters of inquiry and formal proposals. These materials were simple and adaptable; they included a short, introductory letter, a mission statement, a summary of program activities, a statistical description of the numbers of teachers and students participating in the program, and a basic budget.

SUBMIT GRANT PROPOSALS

We submitted these materials to our prospects. Often we used a simple inquiry letter as an introduction, followed by phone calls. We made personal visits to program officers whenever such visits were appropriate, and we worked the telephones when they weren't. We invited grantmakers to meet with representatives of the program, including senior campus officials and the superintendent of the San Francisco Unified School District. We emphasized personal contacts and tried to get grantmakers to meet with teachers, volunteers and student participants and to see the program in action.

MAINTAIN GOOD RELATIONS WITH DONORS

We try to be honest in our reports to donors, describing the successes of the program but not hiding areas that may be problematic. We acknowledge program donors in many ways. We invite them to events, mail them newsletters and T-shirts, and publicly thank them for their support. We submit grant reports on time and in accordance with the funder's procedures. We resolicit each year to sustain their support. Remember one of the most basic principles of fundraising: your best prospects are your current donors.

I include below some additional important considerations:

- Fundraising in support of K-12 education has become intensely competitive and increasingly segmented. Many corporate funders have announced public initiatives to respond to the crisis in America's classrooms. These companies have a legitimate self-interest since many of them require an educated labor force and a reasonably well-informed consumer. You should be aware, however, that many of these companies have identified a niche for themselves. Either they want to fund programs in specific disciplines or they may want to tie themselves to a particular location through some form of "adopt-a-school" program. Many of these efforts have merit and are well-conceived and well-executed, but they may have no connection to what you are trying to do. Save yourself time and effort by finding out what the funder's objectives are before you submit a letter of inquiry or schedule an appointment.

  Avoid the temptation to change your program just to fit a grant. Also do not write that you are going to do something if you have no intention of doing it but are just writing it to meet the donor's guidelines. If you are in doubt whether what you want to do fits the guidelines, talk to the foundation program officer. Be honest. Their guidelines may not be as rigid as they sound.

- Federal and state governments provide hundreds of millions of dollars in support of precollege science education programs. Virtually all of the major science educational partnerships receive substantial public funding. Information and tips on approaching federal and state sources are included in the article "Raising Money from the Public Sector."

- You MUST have the commitment and support not only of the senior administrative leadership of your institution but also of the superintendent and/or the principal whose district or school your program is designed to serve. Most funders will demand evidence of understanding and enthusiasm on the part of the leadership of both the university and the school. If you cannot produce personal commitments from these people, then your efforts to raise funds (or even have a successful program) are probably doomed. Senior administrators must be willing to meet a few key people, to talk on the phone with other prospects, and to sign letters of support. It is a good idea to have a standard letter of support that can be modified easily and signed quickly.

- If you can, identify the funder's decision-maker and focus your efforts on that individual. It could be a program officer; it could be a board member; it could be a senior corporate officer. Remember another of fundraising's most basic principles: people give money to people. If you can find a decision-maker, you have a better chance of establishing the kind of personal relationship that leads to success.

- If a development officer from the institution has primary responsibility for coordinating fundraising for your program, that person needs access to the leaders of the program and to all information regarding the program. No one can be expected to raise money effectively with partial information and limited access. Somebody involved in the program on a daily basis should work with the development office. It could be that combinations of different people are best suited for different kinds of proposals.

- Identify attractive components that particular donors may want to fund. The hardest money to get is the money that you need most — core support for staff salaries and for office expenditures. You can often get restricted funds to do specialized things such as paying for student interns to work in laboratories, or buying curriculum materials related to a particular topic such as AIDS education. If possible, when soliciting that kind of money, include line items for the staffing and office expenses that are needed to make the restricted use possible. Somebody has to interview the student interns, find the laboratories for them, process the paperwork, help solve problems, provide enrichment activities, and do the necessary follow-up.

- It is never easy to decide how much money to request from a particular donor. Look at what the donor has provided for similar purposes in the past and take into account the donor's normal range of gifts. Sometimes it may be worth asking for a smaller grant in order to increase your odds. Having established a relationship with the donor, you may be able to receive more substantial contributions after you have come to be known and loved. On the other hand, do not be afraid to ask for a large donation when the circumstances are right. There is an advantage, of course, to having multiple donors providing support. If one funder leaves, you are not devastated.

- You will probably have to demonstrate significant contributions being made by the university and the schools to the program. Many donors expect to see actual financial commitments as well as "in-kind" contributions from both sides of the partnership. This can be particularly difficult for the precollege partner during this period of teacher lay-offs and severe budgetary constraints. Keep in mind also that funders are interested in providing seed money but are unwilling to have a program become dependent on them for its very existence. Therefore, you must try to institutionalize your program as much as possible.

It is worth trying to convince your institution to commit progressively more money to the program. This will reduce the amount of external support that you need to raise year after year. It also makes it easier to raise the money because donors are impressed by programs that have substantial internal contributions. If you are sensing something of a vicious circle being described here, your antennae are very acute. There are no easy solutions to this dilemma.

- The most powerful advocates for your program will be the students that you are serving. Classroom teachers and the program's staffers can also be useful as spokespeople. Nothing is as effective with a prospective donor, however, as direct exposure to articulate, enthusiastic young people who are benefiting from your program.

There are many compelling reasons for institutions of higher education to offer their resources in collaborations with precollege school systems. Responding to the financial realities of these arrangements usually requires the participation of the private sector as well. Such partnerships among schools, universities and private funders offer rich rewards for all involved.

Good luck.

THE GRANT DONOR/RECIPIENT PARTNERSHIP

BY KATHRYN CAREY

Aristotle said, "To give away money is an easy matter and in any man's power. But to decide to whom to give it and how much and when, and for what purpose and how, is neither in every man's power nor an easy matter."

The American Honda Foundation encounters Aristotle's challenge on a daily basis. Like all foundations, we receive many more solicitations than we can possibly fund. The foundation people who have to make the difficult decisions are a board of directors and a foundation staff. The staff, consisting of a manager, an administrator and a senior program officer, is responsible for the day-to-day operation of the foundation. The board of directors makes the final decision on grant awards based on the information assembled and analyzed by the staff.

At American Honda, "who you know" is not relevant. We want to see "what you know" about solving problems and providing solutions to complex societal issues. Our goal is to select grant proposals that accomplish the most good as parts of structured programs — programs that balance the most pressing needs of youth and scientific education against the available foundation funds. To even be considered, a proposal must have national impact or implication, and it must be explicitly related to youth and science education. The submitting organization also must have the nonprofit 501(c)(3) status from the IRS.

We have a systematic approach to rating the hundreds of worthy grant solicitations that we receive. The scoring process that we use is unique to Honda at this time, but many other grantmakers have inquired about the process and are considering implementing it. The Honda Foundation staff and board of directors have established 13 "Want" Objectives. Each of these categories is assigned a weight signifying its importance on a scale of 1 to 10, with 10 being the maximum importance. These "Want" Objectives are:

Scientific	10
Youthful	10
Broad in scope	9
Soundly managed	8
Financially sound	8
Potential for success	7
Foresightful (forward thinking)	7
Degree of duplication	6
Dreamful (Imaginative)	5
Creative	4
Humanistic	3
Urgency	2
Risk	1

For each grant proposal, the foundation staff assigns a score from 1 to 10 for each "Want" Objective. That score is then multiplied by the weight of that category and the totals for all the "Want" Objectives yields a final weighted score for each and every proposal. A similar weighted scale enables us to quantify possible adverse consequences such as failure of the program over time, misuse of program funds, and lack of public support. Each grant proposal emerges from this process with an overall point score. We believe that this process enables us to meet Aristotle's challenge as objectively, rationally and democratically as possible.

The top scoring 10% of proposals advance to the next evaluation step — the site visit. The American Honda Foundation does not award a grant without first making a site evaluation. A site evaluation invariably provides a better and more complete perception of how the organization actually functions. In the words of Arnold Shore, formerly from the Exxon Education Foundation, "A visit adds flesh to guesses."

A grant invests in people, not in words on paper or a glossy brochure.

The people in charge can either make a project work or cause its failure. Meeting those people on a one-to-one basis and seeing their program in operation enables the foundation to ascertain the program staff's ability, commitment and enthusiasm. Sometimes what appears to be a mundane program on paper turns out to actually be an exciting, worthwhile project due to unique staff resources. The converse is true on occasion as well. A program that looks excellent on paper may have been composed by a professional grant writer. In reality, the program does not stand up under personal scrutiny.

Corporations like to see active, inspired, participatory projects. After all, people are the lifeblood of the corporation. Corporations like people that operate with determination, confidence and authority. Visiting sites is perhaps the most rewarding activity for foundation staff. In a world where headlines blare of recession, gang murders, staggering dropout rates and homelessness, it is both uplifting and exciting to experience the tireless efforts of dedicated people to improve the quality of life for all members of society.

Successful programs result from a successful partnership between the grant recipient and the grant donor. Grantmakers often request that applicants jump through hoops, be faster than a speeding bullet and leap tall buildings in a single bound — all in the name of impressing the board of directors. Grantseeking and grantmaking should be a partnership: an easy, friendly, flexible relationship that allows for change, and perhaps even anticipates it. This partnership must be based on certain intangibles offered willingly and freely by both grantseekers and grantmakers, intangibles such as trust, honesty, sincerity and openness.

Grantseeking and grantmaking should be a partnership: an easy, friendly, flexible relationship that allows for change, and perhaps even anticipates it.

One important issue that these partnerships must face is the ability of the nonprofit organization to continue operating successful programs beyond the initial grant period. Very few foundations, particularly corporate foundations, fund operating expenses. Why? Traditionally, because it is not glamorous to do so. It's not considered to be new, innovative, creative or "cutting edge" work.

Many foundations want nonprofits to develop a brand new project. These projects are then typically funded by the donor for a few years, usually as a demonstration or "seed" project. But then what? Usually, the foundation elects not to continue the funding, and the nonprofit is saddled with yet another project, which may be outside the primary scope and purpose of the organization. It is thereby forced to seek scarce operating funds to continue the project, if it is continued at all.

Is this type of giving the true meaning and purpose of philanthropy? The nonprofit is stuck with a program that is no longer perceived in the grant world as new or glamorous. In the worst-case scenario, the nonprofit will drop the program altogether, and move on to the next funding partner and the next new, "cutting edge" project. How can these kinds of short-term funding cycles cause the positive changes that both grant donors and grant recipients want to see as the results of their work?

Many "old" programs work. They worked when they were first implemented. Over time, any bugs have been ironed out, and the program now runs along smoothly, servicing its designated constituency. What is wrong with funding a program like this? Nothing! It is time-worn and proven. To survive, it requires general operating funds — nothing fancy, just money to go from day to day, month to month, year to year. Good ideas, good programs stand the test of time. It may not be glamorous . . . it may not be cutting edge . . . but it is good work.

In a world where many excellent programs require continuing funds and where worthy new ideas need seed money, the staff of a corporate foundation must say "no" far more often than "yes." Sheer numbers of proposals received compared against dollars available necessitate this posture. It's one of the biggest frustrations of the job. But, on the other hand, corporate foundation staff who desire to make a difference see the fruits of their labors grow through partnerships with nonprofits. At American Honda, we know that it is possible to meet the demands of our business as well as the challenges of society.

RAISING MONEY FROM THE PUBLIC SECTOR

BY ART SUSSMAN

The federal fiscal year 1993 budget request included $768 million for precollege math, science and technology education programs. The figure for 1990 was $344 million, less than half the 1993 amount. Given the relatively large current allocation and the potential for continued increases, the public sector is a promising option for supporting science education partnerships. This conclusion also follows from the U.S. government's continued emphasis on the fourth National Education Goal to make American students "first in the world" in science and mathematics achievement.

This article is an introduction to federal and state sources for funding science education partnerships and suggests the best approaches to take. The companion article by Dennis Hartzell, "Raising Money from the Private Sector," has many suggestions that are equally applicable to the public sector. The ones that are particularly appropriate here include:

• Target applications to the most appropriate sources.

• Do not fundamentally change your program just to fit a grant.

• Obtain partnership commitment at the highest organizational levels.

• Provide as much local and in-kind support as possible.

• Identify components that are attractive to donors.

Unfortunately, the federal government does not make it easy to identify all the potential sources of funding for your science education partnership program. The situation has improved in recent years with the establishment of the Federal Coordinating Council for Science, Engineering and Technology (FCCSET, pronounced "fix it"). The

FCCSET Committee on Education and Human Resources (CEHR) has provided annual reports documenting the federal funding that is targeted specifically for math, science, technology and science literacy education programs. The table included here details the fiscal year 1993 budget request by agency and major program area. The amounts are in millions of dollars per year.

Clearly, the two most important sources of federal funds are the Education Department (ED) and the National Science Foundation (NSF). Since most partnership programs focus on teacher staff development, the Department of Energy (DOE) and NASA stand out among the other agencies. Funds from DOE are generally associated with DOE sites such as Argonne National Laboratory, Brookhaven National Laboratory and Lawrence Berkeley National Laboratory.

With a total budget exceeding $30 billion and more than $3 billion available in competitive grants, the ED is clearly the most influential federal player in the field of precollege and postsecondary education. Since the department has numerous programs, a wealth of acronyms, and an ever-changing funding situation, knowing which programs to target in ED is not a simple matter. What follows is only an overview of relevant 1993 ED programs. Resources described later in this article

FY 93 Budget Request by Agency and Program Area

Major Category	TOTAL	DOD	ED	DOE	HHS	DOI	NSF	NASA	EPA
Precollege Total	768	5.0	371	32	22	25	286	17	8
Teacher Preparation and Enhancement	437	0.6	287	17	5	2	115	8	2
Curriculum Development	92		3	5	3	7	68	5	2
Comprehensive and Organizational Reform	104		20	4	4		74		2
Student Incentives	68	4.4	20	7	10	14	11	1	
Program Evaluation and Studies, and Dissemination	59		42				14		1
Other	9					2	5	3	

Key to acronyms: DOD = Dept. of Defense; ED = Education Dept.; DOE = Dept. of Energy; HHS = Health & Human Services; DOI = Dept. of Interior; NSF = National Science Foundation; NASA = National Aeronautics and Space Administration; EPA = Environmental Protection Agency will help you to obtain up-to-date information.

ED's Dwight D. Eisenhower Mathematics and Science Education program is the largest math and science teacher enhancement program. Often called Ike, the Eisenhower program has several components. All but 5% of the Eisenhower allocation goes to the 50 states where it is then divided into two pools. The larger of the pools of money (75%) is distributed to local school districts (LEAs, Local Education Agencies) to support math and science teacher training. In a typical district, this money is housed in the office of the science/math supervisor, the superintendent, or the grants administrator. LEA Ike money can be used as in-kind support to pay for teacher participation in a science education partnership program. However, do not be surprised if the local school district is reluctant to "share" this money as it often is the only discretionary money that the district has to support science and mathematics education.

Every school district is eligible to receive LEA Ike funds. All a district has to do is submit a fairly simple application to the state. The amount of funding is distributed according to a formula that takes into account the number of schools where more than half of the students enrolled come from low-income homes. Every school district is entitled to receive a base funding and sometimes small school districts pool their Ike money to have a reasonable-size program. In recent years, the LEA Ike program has been criticized for providing inservice opportunities that may not be long enough to have a significant impact. Future LEA Ike allocations may emphasize teacher training that lasts two weeks or more.

The other 25% of state Ike funds are awarded on a competitive basis to Institutions of Higher Education (IHEs) that form partnerships with one or more elementary or secondary districts or schools. These partnerships also can involve the business community and informal science education centers. The Ike higher education funds can be used to provide teacher staff development in science and math and for programs that directly benefit underserved students in these subjects. The State Eisenhower Higher Education funds are an excellent source for science education partnerships. For more information, contact your state Office of Education to get information about the State Higher Education Eisenhower program. Each state has a Higher Education Eisenhower coordinator. Information also can be obtained from the

Dwight D. Eisenhower Mathematics and Science Education Program; U.S. Department of Education; 400 Maryland Avenue, SW; Washington, DC 20202.

ED has other programs that support mathematics and science. For example, the National Clearinghouse and the Regional Consortia have a math and science focus but are not sources of funding for partnerships. Instead, they can provide resources that are useful for partnerships. These are described in the article "Resources for Science Education Partnerships." Most of the other ED programs that can support precollege science education do not specifically target science but generally encourage reform and innovation in elementary and secondary schools. Currently, these include the Educational Partnerships Program, the Fund for the Improvement and Reform of Schools and Teaching (FIRST) and the Fund for Innovation in Education, all three housed in the ED Office of Educational Research and Improvement (OERI; 555 New Jersey Avenue; Washington, DC 20208).

The ED Office of Postsecondary Education has programs that address general curricular and reform issues, but also can be applied to science education. Perhaps the best known is the Fund for the Improvement of Postsecondary Education (FIPSE). Approximately one-quarter of FIPSE projects in recent years have dealt with mathematics and science in four broad categories — curricula, staff development, minority access/retention and preservice teacher education.

ED also features many programs that target special needs, such as programs for Native Americans, the gifted and talented, physically challenged students and bilingual populations. The availability, focus and priorities for these programs vary in different years and regions. If your location or expertise addresses a specific issue of national interest, you may be able to receive funding through programs that target that particular issue rather than through a science focus. Your content focus also may fit opportunities in other federal agencies, such as NASA, EPA or Health and Human Services.

The National Science Foundation is the other federal agency that provides significant support for science education partnerships. For example, more than 70 programs in California received NSF teacher enhancement support in 1992. The Directorate for Education and

Human Resources (EHR) is the branch of NSF that is of greatest interest for science education partnerships. EHR has a number of different divisions that are relevant here.

The EHR Division of Elementary, Secondary and Informal Science Education supports activities designed to develop, improve and disseminate effective instructional materials and methods in both informal arenas and classroom settings and to improve the qualifications and effectiveness of science, engineering and mathematics elementary and secondary school teachers. Its primary programs are: Teacher Enhancement, Instructional Materials Development, Informal Science Education, Presidential Awards for Excellence in Science and Mathematics Teaching and Young Scholars.

The EHR Division of Undergraduate Education houses the Teacher Preparation program. This area was previously linked to Teacher Enhancement but was moved to the Undergraduate Division because that is where prospective teachers receive their training. NSF hopes to connect reform in teacher preparation with reform in undergraduate science education. Collaborations among scientists, science educators, teachers and other educational leaders, within and among institutions of higher education and school systems, are encouraged.

The EHR Division of Human Resource Development has primary responsibility for broadening participation of underrepresented groups in science and engineering. At the precollege level, three programs are of particular interest: Comprehensive Regional Centers for Minorities, Summer Science Camps and Partnerships for Minority Student Achievement.

The EHR Office of Systemic Reform manages and operates broad-based initiatives to stimulate major reform in precollege science and math education. Currently, 25 states and Puerto Rico have received State Systemic Initiative funding of $2 million per year for five years. A new Urban Systemic Initiative program will target cities with large populations of low-income minorities. Science education partnership programs should attempt to align their programs with these efforts, as well as with state efforts in curriculum framework and assessment reform.

The EHR Division of Research, Evaluation and Dissemination serves all EHR units by supporting data collection, conducting analytic and evaluative studies, and evaluating individual projects and programs. It also supports the dissemination of the results of NSF-supported projects. The Applications of Advanced Technologies unit within this division funds research on the applications of advanced technologies, particularly the computer, to science and mathematics education.

NSF has a variety of publications that describe the programs and application procedures. The "Guide to Programs, FY 93" (publication NSF 92-78) is available from NSF Forms and Publications Unit; 1800 G Street NW, Room 232; Washington, DC 20550. NSF also has an on-line publishing service called STIS, the Science and Technology Information System. Many NSF publications can be accessed and downloaded through STIS. To access the system, follow the instructions on the STIS flyer (NSF 91-10, rev. 10-25-92 from the Publications Unit). To get an electronic copy of the flyer, send an e-mail message to: stisfly@nsf.gov (Internet) or stisfly@NSF (BITNET).

NSF often requires a preproposal six weeks before the official application submission date. NSF program officers like to provide extensive feedback throughout the proposal preparation and review process.

This information on the DE and NSF is just a beginning. There are several complementary approaches to expand your search for the most appropriate public sector funding sources. The Catalog of Federal Domestic Assistance, published by the U.S. Office of Management and Budget, describes all federal programs that distribute funds to states, organizations and individuals. The domestic subscription price is $38 annually and it includes periodic updated materials. It is available in most major libraries or by subscription from the Superintendent of Documents; Government Printing Office; Washington, DC 20402.

The Federal Register is a publication that is issued every weekday listing all federal agency regulations and legal notices, including details of all federal grants competitions. It is available in most major libraries or by subscription ($170 for six months by second-class mail) from the Superintendent of Documents. The Federal Register often provides the first official notification of a new grant program.

Perusing these publications that list all government programs, not just those relating to education, can be time-consuming and frustrating. One alternative is to subscribe to a commercial newsletter service that specializes in education. The "Education Grants Alert" provides weekly reports on funding opportunities for K-12 programs. It costs several hundred dollars per year and is available from Capitol Publications; P.O. Box 1453; Alexandria, VA 22313. The Dwight D. Eisenhower Mathematics and Science Education Newsletter is free, but considerably less comprehensive. This quarterly newsletter is available from Consortium for Educational Equity; Rutgers University; Building 4090, Livingston Campus; New Brunswick, NJ 08903.

Technology may be coming to the rescue. The National Network of Eisenhower Mathematics and Science Regional Consortia and National Clearinghouse is developing an on-line database to serve as a central source of information on all federal programs that support K-12 science and mathematics education. There also will be publications targeted to each of 10 geographical regions. This database will go on-line in spring 1994, with hard-copy publications available by January 1994. Addresses and phone numbers for the National Clearinghouse and the Regional Consortia are provided in the article "Resources for Science Education Partnerships."

Another way to stay informed is to get involved in science education reform efforts in your state. Attend state, regional and national science education conferences, such as those sponsored by the National Science Teachers Association and local affiliates. Get to know coordinators of other partnerships, join professional organizations, trade information and call government agencies to track down rumors. Get your name on appropriate mailing lists. However, a word of caution with respect to the federal government: Do not rely on having your name on a mailing list as your primary source of information about programs.

Be prepared to put together a proposal on short notice. Keep those wonderful phrases in your computer ready for cutting and pasting. And do not get too discouraged by rejection letters. See if there are opportunities to adjust and resubmit your proposal. Even though funding for precollege science education programs has been growing, there are still many more good programs seeking funds than money available to support them. Our best hope for realizing the improvements that we all want to see is to somehow convince our nation to allocate the sustained resources that the task demands. Then, instead of devoting so much talent and energy to securing scarce dollars, we could focus more on the educational work at hand.

RESOURCES FOR SCIENCE EDUCATION PARTNERSHIPS

BY ART SUSSMAN

Y̶ou are not alone. The variety of articles and authors in this book indicates that there are lots of us out there creating many kinds of fruitful interactions between schools and science-rich institutions. The more we communicate with each other, the more we can profit from the lessons that we have learned, and the greater the force that we can generate to create a better educational system.

Far from being alone, a large network exists that can provide significant resources as we work together to foster systemic change. Some of these resources have focused on precollege science education for many years. Others are powerful groups, such as the National Academy of Sciences, that only recently have begun to play a significant role.

This chapter provides an overview of the kinds of organizations and services that are available. A detailed listing would not only double the size of this book, it would be a needless duplication of effort. The American Association for the Advancement of Science (AAAS) publishes an annual *Sourcebook for Science, Mathematics, & Technology Education.* This invaluable reference book can be obtained from the Directorate for Education and Human Resources, AAAS, 1333 H Street NW, Washington, DC 20005 (202-326-6670).

AAAS is one of the most important kinds of resources, the professional organizations. Founded in 1848, it is the world's largest federation of scientific and engineering societies. It has approximately 300 affiliated societies and more than 130,000 members. AAAS sponsors numerous programs and many of its affiliate societies offer their own more specialized resources for precollege science education. The AAAS *Sourcebook* is the best detailed guide to the offerings of AAAS, its affiliates, and all the other kinds of resources that we will mention in this article.

AAAS precollege programs are typical of the resources provided by associations. AAAS publishes a newsletter, a review journal of science materials, the *Sourcebook*, and analyses of issues in precollege science education. It sponsors programs to connect scientists with schools, teachers and students. One example is Science-By-Mail, a pen pal project that directly links scientists with students in grades 4-9 who are investigating three assigned challenge packets. AAAS sponsors a variety of programs to work with schools and community organizations to increase the participation of women and underrepresented minorities in science and mathematics courses and careers. It also has developed curriculum materials. Its most ambitious effort, Project 2061, is a multiyear initiative to define elements in science and mathematics education that are the substance of science literacy. Also typical of large associations, many of the societies affiliated with AAAS have their own, more specialized precollege programs.

The National Science Teachers Association (NSTA) is another major association that provides valuable resources for science education partnerships. The world's largest organization of science teachers, NSTA networks science educators nationwide through 50 state chapters with journals, newsletters, conferences and an electronic bulletin board, "Science Line" (202-226-4496). More than 25,000 educators, scientists and vendors attend one national and three area conventions annually that feature lectures, workshops, seminars, forums, papers and exhibits. NSTA publishes four science teaching journals, one for each major level of science education: *Science and Children* (K-8), *Science Scope* (5-9), *The Science Teacher* (7-12) and the *Journal of College Science Teaching*. Members also receive *NSTA Reports!*, a science education newspaper published six times a year.

NSTA's involvement in science education includes much more than conferences, journals and newsletters. At the reform level, NSTA has led a national Scope, Sequence and Coordination (SS&C) reform effort that seeks to integrate the teaching of different science disciplines and replace the current fragmented, layercake approach (biology, chemistry, earth science, and physics taught in separate courses). With the motto, "Every Science Every Year for Every Student," SS&C offers an alternative to the current system in which many students never understand the connections among the fields of science and do not learn chemistry or physics.

As a professional organization, NSTA also has programs for certification of science teachers at the different grade levels as well as in discipline area categories. NSTA issues policy statements on important science education issues, such as equity, laboratory safety, use of animals in classrooms and assessment. The organization also provides access to more than 25 award programs that provide prizes, recognition and/or institutes. For membership and general information, contact NSTA at 1742 Connecticut Avenue, NW, Washington, DC 20009 (202-328-5800).

Numerous professional organizations provide services that target specific disciplines or issues. The AAAS *Sourcebook* gives the addresses and brief information about hundreds of these groups. If a particular focus applies to your partnership activity, one of these numerous organizations can provide you with valuable curricula, contacts, career information or ideas. Examples of these kinds of associations include: American Chemical Society, American Dental Association, American Society for Microbiology, Association for Women in Science, Geological Society of America, National Association of Biology Teachers, National Science Supervisors Association, National Society of Black Engineers, Optical Society of America, Society for the Advancement of Chicanos and Native Americans in Science and The Wildlife Society.

A few organizations focus on education partnerships. The three most prominent are the National Association of Partners in Education, Inc. (NAPE), the Triangle Coalition for Science and Technology and the American Association for Higher Education (AAHE). NAPE is oriented more toward business-school partnerships and is located at 209 Madison Street, Suite 401, Alexandria, VA 22314 (703-836-4880). The Triangle Coalition has a variety of programs and resources to support science alliances and to influence government policies and science education funding. It is located at 5112 Berwyn Road, 3rd Floor, College Park, MD 20740 (301-220-0885). AAHE has established a School/College Trust with nine staff members to promote school/college partnerships and sponsor an annual conference. Recently, AAHE has expanded its agenda by promoting systemic reform at the college level linked with K-12 reform. For more information, write to AAHE School/College Trust, American Association for Higher Education, One Dupont Circle, Suite 360, Washington, DC 20036 (202-293-6440).

Federal and state government agencies form another very important class of resource providers for science education partnerships. The article, "Raising Money from the Public Sector," highlights the role of government in awarding that greatly treasured resource, money, the food of all good. But government agencies can provide other resources. A State Department of Education will typically house several to many capable administrators who can provide information about local programs and resources. These include science supervisors, mathematics supervisors, assessment specialists, coordinators for the state Eisenhower science and mathematics programs, and leaders of statewide systemic reform initiatives. At the state level, there are also academies of science that have aims and objectives similar to those of AAAS. The AAAS *Sourcebook* lists the names of these state education administrators and science academies.

At the federal level, the main providers of non-monetary resources are the Department of Education, the Department of Energy, and the National Science Foundation. If you are interested in a particular topic, it is well worth contacting a federal agency to discover if it has inexpensive or free background materials that can be obtained directly from the agency or via the U.S. Government Printing Office. For example, the Energy Information Administration (202-586-8800) of the U.S. Department of Energy offers various services and information packets. Other examples are the many National Institutes of Health (such as the National Institute of Mental Health) of the Public Health Service, NASA, the Environmental Protection Agency and the Department of Agriculture.

The Eisenhower National Clearinghouse for Mathematics and Science Education (ENC), funded by the U.S. Department of Education, is a new access avenue for math and science resources. Located at Ohio State University, ENC received its first funding in October 1992. The clearinghouse will collect and create the most up-to-date and comprehensive listing of mathematics and science curriculum resources in the nation. The list or catalog of materials, the text of some of the materials, and evaluations of them will be made available in a database in a variety of formats, including print, CD-ROM, and electronically online. To receive the ENC newsletter and brochures, write to Eisenhower National Clearinghouse for Mathematics and Science Education, Area 200, Research Center, 1314 Kinnear Road, Columbus, Ohio 43212 (614-292-7784).

Ten Regional Consortia for science and mathematics education were funded at the same time as the Eisenhower National Clearinghouse. Together with ENC, they form a national network to promote systemic improvement. Each Regional Consortium serves a contiguous multistate region by providing technical assistance at the state and local levels; networking assistance at the local, state and regional levels; and identification and dissemination of exemplary programs. The Regional Consortium serving your area should be a good source of information about the ENC, activities and key players in your state and region, exemplary programs and practices, electronic networks, and policies in your state and region with respect to new standards, revised frameworks and reformed assessments.

As mentioned in the article, "Raising Money from the Public Sector," the National Network of Eisenhower Mathematics and Science Regional Consortia and National Clearinghouse is developing an on-line database to serve as a central source of information on all federal programs that support K-12 science and mathematics education. There will also be publications targeted to each of 10 geographical regions. This database will go on-line in the spring of 1994, but the publications should be available by January 1994. The address for each Regional Consortium is provided on the next page.

The National Diffusion Network (NDN) is another program that can help partnerships locate and implement exemplary curriculum materials. Administered by the U.S. Department of Education, NDN assists schools in adopting approved programs. NDN programs have been field-tested, evaluated locally and approved by a panel of the U.S. Department of Education. An NDN State Facilitator serves as the in-state link with the national program. For more information, contact National Diffusion Network, U.S. Department of Education, 555 New Jersey Avenue NW, Washington, DC 20208, (202-219-2134).

Region: Northeast
States served: ME, NH, VT, MA, RI, CT, NY
Northeast Regional Alliance for Mathematics and Science Education Reform
300 Brickstone Square, Suite 900
Andover, MA 01810
508-470-0098   Fax: 508-475-9220

Region: Mid-Atlantic
States served: PA, NJ, DE, MD, DC
Mid-Atlantic Regional Consortium for Mathematics and Science Education
444 North Third Street
Philadelphia, PA 19123
215-574-9300   Fax: 215-574-0133

Region: Appalachia
States served: VA, WV, KY, TN
Eisenhower Math/Science Consortium
P.O. Box 1348
Charleston, WV 25325
304-347-0400   Fax: 304-347-0487

Region: Southeast
States served: NC, SC, GA, FL, AL, MS
SERVE Mathematics and Science Regional Consortium
345 South Magnolia Drive, Suite D-23
Tallahassee, FL 32301
904-922-2300   Fax: 904-922-2286

Region: North Central
States served: OH, MI, IN, WI, IL, MN, IA
Midwest Consortium for Mathematics and Science Education
1900 Spring Road, Suite 300
Oakbrook, IL 60521
708-571-4700   Fax: 708-571-4716

Region: Mid-Continent
States served: ND, SD, NE, WY, CO, KS, MO
High Plains Consortium for Mathematics and Science
2550 South Parker Road, Suite 500
Aurora, CO 80014
303-337-0990   Fax: 303-337-3005

Region: Southwest
States served: AR, LA, OK, TX, NM
Southwest Consortium for the Improvement of Mathematics and Science Teaching
211 East Seventh Street
Austin, TX 78701
512-476-6861   Fax: 512-476-2286

Region: Northwest
States served: MT, ID, WA, OR, AK
Northwest Consortium for Mathematics and Science Teaching
101 South Main Street, Suite 500
Portland, OR 97204
503-275-9594   Fax: 503-275-9489

Region: Far West
States served: UT, AZ, NV, CA
Far West Regional Consortium for Mathematics and Science Education
730 Harrison Street
San Francisco, CA 94107
415-241-2730   Fax: 415-241-2746

Region: Pacific
Service area: HI, American Samoa, Northern Mariana Islands, Palau, Marshall Islands, Micronesia Pacific Region Mathematics/Science Consortium
1164 Bishop Street, Suite 1409
Honolulu, HI 96813
808-532-1900   Fax: 808-532-1922

Scientists and teachers located near a facility of the U.S. Department of Energy (DOE) can benefit from the department's commitment to supporting precollege science education. These DOE facilities include multiprogram laboratories (such as Lawrence Berkeley Lab, Argonne National Lab, National Renewable Energy Lab, Brookhaven National Lab, Los Alamos National Lab), program-dedicated facilities (such as Fermilab, Nevada Test Site, Stanford Linear Accelerator), and university consortia (such as Associated Western Universities). These facilities support workshops for teachers, partnerships with scientists, programs for students, and school or district-wide reform efforts. A *Catalog of Education Programs* is available from United States Department of Energy, Office of Scientific and Technical Information, P.O. Box 62, Oak Ridge, TN 37831.

The National Academy of Sciences (NAS) has begun to play a prominent role in the science education reform movement. The development of national science education standards is being coordinated by the NAS. The NAS involvement began with the establishment of a National Science Resources Center (NSRC) in conjunction with the Smithsonian Institution. The NSRC has been most heavily involved in elementary science education and in promoting local resource centers to distribute hands-on science supplies to teachers. An NSRC book, *Science For Children: Resources for Teachers*, is available for $7.95 from National Academy Press, 2101 Constitution Avenue NW, Washington, DC 20418.

Back at the local level, most science and technology centers that have programs for the public include a variety of resources and programs for students and teachers. These institutions include science museums, zoos, planetariums, nature centers, natural history museums and aquariums. The AAAS *Sourcebook* lists many but not all of these organizations. It is well worth contacting any local organization or center that provides exhibits, information or public attractions that have a scientific or technological basis to discover if they have resources that you can use in your science education partnership activities.

One of the six National Education Goals is for American students to excel in mathematics and science. The many resource providers cited here can help scientists and teachers work together to achieve this goal for all students.

Dave Olson, a high school computer science teacher, and his industry mentor, Steve Belochi, examine a big picture on the computer screen (photo courtesy of IISME).

(Right) Teacher Joan Regan analyzes mathematical relationships on a globe as part of The Exploratorium's Science at the Core staff development program (photo courtesy of The Exploratorium).

THE BIG PICTURE

To Change Our Cities

Perspectives on Systemic Change

Ameristroika

TO CHANGE OUR CITIES

BY LEON M. LEDERMAN

I am chronologically advantaged enough to recall a presidential candidate in the 1930s tell us: "One third of a nation is ill-housed, ill-clothed and ill-fed." Of course, this was Franklin Roosevelt, and he was talking about the underclass in 1930. So many things have happened since then.

As a nation, we have grown immensely more affluent by any measure you may choose. We have had long periods of intense concern about the poor, the underclass, minorities; we have legislated equality, and ruled against segregation in Brown vs. Board of Education. We have an extensive welfare system, some Head Start programs, and other public and private initiatives too numerous to know. But the 1930s assessment is still with us. Although only a fool would say we have not made any progress, that one third of a nation living in poverty seems to be an invariant.

An adage I read in a business brochure calls for the courage to change the things that can be changed, the serenity to accept those that cannot be, and the wisdom to know the difference. The persistence of an impoverished class would seem to call for wisdom and serenity to conform to the adage, but it is not always wisdom that shapes what one does. It wasn't wisdom that got me into the campaign I will tell you about. Frankly, I'm not sure I know what it was. Perhaps it is arrogance, the arrogance of the scientist who has had some success in management, or perhaps it is just unreasoned anger. I find myself being very angry these days, but I will try to control this and tell you what I know. It has to do with education. It has to do with public schools in large cities.

First, I must tell you about my city — Chicago. With 410,000 students, it is the third largest school system in the nation. There are somewhat over 20,000 teachers, and 17,000 of them must teach some kind of math and science, largely in primary school. Like so many large cities, the school population is 88% minority, 12% Asian and white. So much for Brown vs. the Board of Education. Over 67% of the children come from families below the poverty level. The dropout rate is officially listed by the Board of Education as 45%, but the more reliable estimate is that 60% to 70% of the children never graduate from high school. For individual schools in the worst districts, the numbers can be almost 90%. Chicago students do very poorly on national tests. Over half the high schools placed in the lowest 1 percentile of ACT scores.

Again, like all other cities in America, the streets and even the school corridors are unsafe — violence, drugs and gangs are part of the territory. Many school buildings are over 100 years old, overcrowded with gyms and corridors pressed into use as classrooms. Jonathan Kozol eloquently documents this appalling situation in his book, *Savage Inequalities*. One significant piece of datum is the increase in the percentage of children born to single women. In 1950, this was 3%; in 1990, it is 28%. What has this to do with education? One of the president's goals is that by the year 2000, all children will be ready for school. What does this mean? It should mean:

- supplemental nutrition programs for pregnant women and young children;

- prenatal care and immunization;

- special attention to the babies born to poor women; and

- encouraging, not discouraging, family planning and sex education.

Children who are part of these statistics have a high probability of being problem children in school: hyperactivity, low attention span, impaired hearing and vision, asthma, and a variety of learning problems resulting from malnutrition and brain damage in utero. Teachers tell us that one or two in a class can be handled without shortchanging the rest, but more than two leads to chaos and paralysis. Ernest Boyer's 1991 study indicates that 35% of American children are not ready for school when they enter kindergarten. Tragically, very little is being done to change the numbers "by the year 2000."

With this as background, let us look again at Chicago, "the worst school system in the nation," according to William Bennett, former secretary of Education. Something interesting happened in Chicago in

1988 — a movement led by outraged parents that was eagerly assisted by university people, the private sector and many other elements of the city. Sharing the view of the public schools as a tragic and no longer tolerable disaster, this coalition combined to pass the most radical school reform law in the nation. Today, at least in principle, Chicago has almost 600 new "corporations," the CEO (principal) is installed by a Board of Directors (local school councils elected by citizens who live

---

The intervention we designed was a massive retraining of the teachers, those 17,000 professionals who must teach math and science and who, for the most part, were never trained to do so.

---

around the school) and he or she must run a successful operation or be dismissed. It is far too early to comment on this reform except for the obvious benefit of stirring up interest by the parents and other citizens — in running for the council, in voting and being, even in a limited way, enfranchised by the reform of the educational system.

After moving to the University of Chicago in 1989, I was drawn to the exciting discussions of how to make school reform work. These discussions gave birth to an idea for how a group of interested parents, teachers, scientists and businessmen could intervene to make a difference. And so was created the Teachers' Academy for Math and Science, which miraculously opened on the campus of Illinois Institute of Technology in mid-Chicago in September 1990.

This private, not-for-profit entity was formally created by a Council of Presidents of all the universities in Chicago. Its board includes teachers, principals, scientists from universities and from two national laboratories, and leaders from the private and public sector. These prominent executives come from Chicago's leading corporations, museums, the Chicago Teachers' Union, Urban League and the Hispanic equivalent, and representatives of the mayor and governor. This is a very large and very complex program. With good will all around, one still has to tread on cross-cultural, political and ethnic eggshells. Yet, it is the only program I know that has the scale to address the president's goal of "being No. 1 by the year 2000."

The intervention we designed was a massive retraining of the teachers, those 17,000 professionals who must teach math and science and who, for the most part, were never trained to do so. Our conviction is that the teacher is the key to a positive, zestful school experience, and to deploying the newest techniques for teaching math and science. Our experience with the teachers has been very positive. We find that they care; they love children; they want to be better teachers; and they hate to be required to teach things that they do not understand. In our view the ultimate tragedy is to have children arrive at school, having overcome whatever obstacles their lives provide, from their bed to the school room, ready to learn, only to discover in their child-wise way, that the teacher is merely and unhappily using time.

Our program involves a hands-on, activity-based curriculum with children working together, talking and doing. Rather than being the authoritative giver of all wisdom, the teacher serves as mentor and guide. These new techniques act as a catalyst, engage the child and the teacher, and are designed to involve the parent or the grandparent. We did not invent this approach. Interventions of this kind have been taking place all over the country — one school here, two schools there, 10, 20, 50 teachers. There were many programs occurring in and around Chicago. The best ones had magical effects. Teachers were re-enthused, while students, engaged by the learn-play activities, were no longer sitting passively at their places, praying that they not be called upon. Now they work together in groups — wrong answers are just as welcome as correct ones — and the objective is the development of thinking skills. Here, through math and science, is a key to true reform of the entire educational process.

Our plan was to "do" the whole city. Exploiting school reform, we would contract with the principal and the local school council. We quickly realized that the scale of our mission required inservice training that would take the teacher out of the classroom while we supplied replacement teachers. Realizing it would evolve over time, we hastily concocted a 16-week intensive training program. We supplemented this with afternoon, weekend and summer programs of follow-up enrichment, technology workshops, museum programs, every possible form of outreach. To get through 17,000 teachers, we planned to do 100 the first year, 1,000 the second year, and 2,500 from then on until year eight, when we could taper back to a maintenance inservice program and a vastly improved preservice process.

Well, what is the reality? We indeed managed to intensively train 109 teachers the first year. An additional 1,500 teachers attended our outreach programs. Now in our second year, we are changing things and we will be lucky if we get through 600 teachers instead of the 1,000 we had initially targeted. We have had over 3,000 teachers attend our workshops and ancillary programs. We intend to have a full-time staff person for one year in all the schools that we cover as one component of what we hope becomes a permanent follow-up program. However, we are frustrated by the difficulties of recruiting capable staff, especially the replacement teachers, and even more by the inexplicable difficulty of securing adequate funding.

Let me summarize the program status and my conclusions. The good news is that it works! Or to state it more carefully, it seems to show promise of working. In addition to built-in pretest and post-test measures, we record comments from teachers, principals and parents as well as our own in-school staff specialists. We measure hours of science and math taught in our schools and in the schools not yet involved, and we have begun to compare truancy rates. In spite of a reasonable number of mistakes, the palpable enthusiasm of teachers and principals is evident. We are concentrating on K-4 now, and next year we will extend from kindergarten through the eighth grade.

If early school education can influence the cycle that traps so many minorities — the cycle of failure and dropout, of poverty and crime, of teen-age pregnancy — then we have, I believe, shown a way to genuine change. The bad news is that it is considered an expensive program. When in full swing, the Chicago effort will need $30 million a year. Under present circumstances, the bulk can come only from the federal government. We were enthusiastically supported by Adm. James Watkins, secretary of the U.S. Department of Energy, and we succeeded in getting additional start-up funds from the National Science Foundation. An early grant from David Hamburg's Carnegie Corporation was enormously helpful. Last year we raised $4 million from federal sources, $2 million from the state, and about $2 million from private sector contributions. This year Illinois, like so many other states, looks hopeless for providing financial support, and Chicago even more so. Floods are also not helpful. Our expectations for federal support have been scaled back from $15 million to about $6 million or $7 million. Support from foundations and businesses continues to be strong, but it takes a huge effort to raise $1 million or $2 million.

From the beginning, our program received the following criticisms: What good is it to fix teachers when the family support is lacking or when learning impairment is already so advanced that a great kindergarten teacher is too late... or when the buildings are so old and crumbling or when...? My response has always been that the critic is indeed correct, but that if we waited for a total solution, nothing would ever start. So we will continue training teachers, and we encourage the critic to get busy on these other things. It seems absolutely clear to me that our society had better get serious about all these issues, and we must do it soon.

Suppose our approach makes sense. In the country as a whole, there are 25 cities like Chicago that could do similar things. We are not marketing the specific programs we have adopted. What we do want to replicate is the marshalling of all the intellectual resources of a community to a purpose. The commitment of university president(s), CEOs of major corporations, scientists and professional educators should be enough to sway a funding agency. Unfortunately, conventional reviews by the professional bureaucrats is still the rule. To extend the kind of thing we are doing to 25 cities and perhaps to intervene in an equal number of poor rural areas would require approximately $1 billion. In Washington, they look at you as if you are really not with it — $1 billion! But is this really too much to invest to provide a lifeline, an escape ladder out of ignorance and poverty?

Yes, there is indeed a fiscal crisis. Yet the Republican president and the Democratic Congress approved a military budget for 1993 that will allocate approximately $140 billion to defend Europe against an attack from the East! Is there any wonder at the irritation of the voters or the frustration of legislators who are quitting, or my own and, hopefully your, personal anger?

My mood alternates. I oscillate from gratitude at getting any help at all (even $6 million is a huge sum of money) all the way to discouragement with the sheer weight of bureaucracy that one must penetrate in order to get anything done. This applies to the federal government, to the state, the city, the school system, even the university. Our teacher enhancement initiative is very often greeted with great praise: "visionary," "bold," "practical" are some of the accolades. Yet, in spite of leaders at the highest level who do grasp the issues, it is also clear that they live under constraints and one still has to prevail way down in the bureaucracies where decisions are made and checks are written. The rhetoric is echoed: "We must break the mold, we must have radical reform." But the real understanding of those words, the real sense of urgency, the awesome implications of an existing teachers corps in which, for example, only 35,000 out of 1 million elementary school teachers (3.5%!) are specially trained for math and science teaching, these are rarely appreciated. No, there is no priority for the underclass, and it creates a most discouraging block to action.

There are many successful interventions. We read and hear about wonderful people who, by the force of their energy and creativity, will turn around an entire school or two. The preschool initiatives are another case in point. But how do we begin a national deployment of what works? How do we pull together the crucial ingredients?

Let me, at the risk of appearing naive, propose something new. Suppose we define a Special Administrator for Cities, charged with the task of bringing the educational achievements of our large inner cities up to the standards we want our nation to have. This program must include teacher retraining, some equipment such as computers, and a large extension of Head Start and the other early intervention programs which are needed to make children ready for school. This special administrator must be educationally expert, nationally known, "street smart" and city-wise. Suppose this special administrator, appointed by the president with the concurrence of key congressional committee chairs, is given a budget appropriation that allows him or her to encourage the formation of consortia of the kind I have described for Chicago. Cities would organize not-for-profit groupings to apply for federal funds. The matching could be 50% federal, and 50% from state, city and private sector contributions. The grants might be for a period of five years, whereupon the federal contribution begins to be reduced as local sources increase.

The administrator could report to a Commission on the Cities, which might include the secretaries of Education, Health and Human Services, and Energy; the directors of the NSF, NIH and NASA; and the president's science advisor and Domestic Council chair. We do not want to pump money through the existing, clogged channels that unfortunately are part of the problem that we need to solve. The size of the administrator's staff should be legally restricted since we want to keep this new (bypass) artery as free-flowing as possible. A proposal by a city should merely be authenticated. Perhaps the president personally asks for and receives a pledge of deep commitment from the president of the local university, the CEO of the corporation, the chair of the city school board, the mayor and governor, etc. Minimize site visits and Washington second guessing. Once a year, the administrator would have the relevant agencies form a visiting committee to gauge and document progress, to ensure fiscal integrity and to improve intercity sharing of experience and data. One could start with three cities and gradually work up to the 25 or so cities that contain a large fraction of the students that we are currently losing.

How much would it cost? Total appropriations for this Cities Initiative might grow to something like $5 billion per year. The increase above the $1 billion amount cited earlier is required to provide the services needed so that students can be ready to learn; $5 billion per year, or even $10 billion, may sound like an impossible figure, but it is really very little when compared to the social and human costs that we pay every day that we do not make these necessary investments in our future. How much do we spend on jails and welfare, on trying to solve seemingly hopeless problems that could be prevented at a fraction of the price? What is the national cost of ignorance, despair and fear? Let us find the courage to change things.

PERSPECTIVES ON SYSTEMIC CHANGE

BY MARK ST. JOHN

In June of 1991, a leading group of funders, "change agents," and mainstream educators met at a three-day conference at Wingspread to focus on science education. Entitled "Science Education for the 1990s: Strategies for Change," the conference was organized, supported and hosted by The Johnson Foundation. This article is drawn primarily from one section of the conference report written by Mark St. John. For a complete copy of that report, please contact Inverness Research Associates; P.O. Box 313; Inverness, CA 94937.

For as long as I can remember, and certainly since the launch of the Sputnik satellite nearly four decades ago, public leaders, the press and the average person on the street have been worried about the quality of the nation's educational system. The concerns are, perhaps, most focused and most urgently felt in the areas of mathematics and science. Funders including the National Science Foundation, the U.S. Department of Education, state legislatures, major corporations and many private foundations have spent billions of dollars, supporting hundreds of projects to help strengthen science and math education, both in and out of schools. In spite of the continuing concern, years of research, and the steady investment of private and public funds, the teaching of science and math practiced in most K-12 classrooms today does not appear significantly different than 20 or even 30 years ago. Certainly, the reality of science and math education in our schools falls well short of the ideals we envision.

The 1991 Wingspread conference brought together mainstream educators, funders and "change agents." These change agents typically have advanced degrees in science or science education and work within institutions that have a wide variety of science resources to offer to the education community. They work at private and public institutions of higher education, technology-based corporations, professional societies of scientists or science educators, informal science education centers, and other nonprofit entities. Authors of articles in this book are change agents.

Conference participants acknowledged that they generally operate within a model of reform based upon projects. Such projects typically are short-term, temporary efforts focused on a particular dimension of change (e.g., curriculum development, assessment, training teachers, etc.). Projects generally provide extra and temporary resources, typically to university-based experts (scientists and educators) to develop innovations and provide training to K-12 practitioners. Funders operate on the basis that it is their role to initiate reforms, but definitely not their role to provide permanent support or continuing funds to cover operating expenses. Moreover, while funders may provide funds for local projects, they usually seek broader, often national, impact through dissemination and replication. The assumption underlying this approach is that specific programs or practices can be validated and transplanted.

Within their context, funders have provided support for projects that have been successful in meeting their own goals. Past projects have produced high-quality and innovative curricula; teachers have been provided with a range of opportunities for developing their knowledge and skills in teaching science; schools are now experimenting with new modes of management; and researchers are testing out new modes of "authentic assessment." Yet, in spite of these efforts, schools and classrooms across the country are, as a whole, largely unchanged. In fact, there may even be decreasing opportunities for many young children to have constructive experiences in science.

There may be two primary reasons for this dilemma. One has to do with the scale and inertia of the educational system relative to the scale of the resources available for improving it. One cannot change a river with an oar. The second reason revolves around the idea that the basic metaphor of change within which funders and change agents operate may not describe well the actual systems they are attempting to influence. The project-centered, infusion-of-innovation metaphor may imply assumptions underlying funders' investments which are not true, and lead them to images of their funds "catalyzing" change — an idea that is naive and overly mechanistic.

A project-centered approach might work when the projects are aimed at contributing to a system which itself is relatively healthy, flexible and capable of improving itself incrementally. The system, in short, must be capable of selecting innovations and incorporating them into daily operations. The system must be capable of self-management in replacing some of its traditional practices. Clearly, this is not the case with science education today.

In fact, all of our experience suggests that one-dimensional, fragmented reform efforts will not work. The problem cannot be divided and conquered by isolated projects. Lessons learned from a single project's success are not "generalizable." Rather, it appears that our reform efforts must be comprehensive in scope and systemic in their execution. Such systemic efforts will have to address simultaneously many dimensions.

In our discussions at the Wingspread conference, several participants wondered about funding in other domains (e.g., health, agriculture, etc.) and in the educational programs of other countries (e.g., England, Japan, etc.). As a result of those questions, I briefly explored some of the lessons learned to date in the area of international investments aimed at "catalyzing" third world development.

It is interesting to note very similar parallels between the reform movement in science education and international efforts to help develop rural communities in the third world. In the 1960s, there was great optimism about projects that would disseminate superior technologies to rural farmers. Much like NSF's innovative curriculum development efforts and its summer institutes in the 1960s, it was assumed that innovations, once proven superior, would be readily disseminated and "adopted" throughout the land. Several parallel issues arise, and each is discussed here in some detail.

Issue: The Adaptability of Innovations

Disillusionment with a project-based diffusion of innovation model has come slowly in the development community. Even in the eyes of the proponents of an adoption paradigm, it was soon recognized that there were critical qualities of an innovation that affected the rate of its adoption:

- Relative advantage — the degree to which an innovation is in fact superior to its replacement. (It must clearly buy you more than it costs in both time and resources.)

- Compatibility — the degree to which an innovation is consistent with existing values and past experiences. (Note that this runs somewhat counter to the degree of its innovativeness.)

- Complexity — the degree to which an innovation is difficult to understand and use. (Technical superiority is useless if the "human interface" is not intuitive to the actual user.)

- Divisibility — the degree to which an innovation can be tried on a limited basis. (This raises the issue of replacement again; few are willing to burn their bridges before much incremental testing.)

- Communicability — the degree to which the innovation and the results of its use can be communicated to others.

Typically, developers launched a campaign to disseminate innovations focusing entirely on achieving maximum "exposure" to the innovation without enough attention paid to the appropriateness of the innovation to the local culture. Innovations tended to have a pro-technology bias — assuming that higher levels of sophistication and complexity in technology would be seen as beneficial. The communication efforts to disseminate these efforts were then, not surprisingly, aimed at the most progressive of farmers.

In many ways the same criticisms can be leveled at the idea of exporting innovations developed in universities and model schools to mainstream schools across the country. Too often, the sociological aspects of innovation are ignored by the innovator who typically is primarily involved in the technical or technological aspects. Failures that seem particularly analogous include the histories of introducing computer technology, videodisks, and other advanced technologies into science education practices. Even today, calculators and hands-on materials for science are often seen as "alien objects" in schools.

ISSUE: INTERNAL BARRIERS

This school of research viewed socio-psychological (cultural) factors as the most severe barrier to innovation. Together these factors comprised "a subculture of peasantry," which included, in part, the following elements:

- Mutual distrust in interpersonal relationships. (Peasants were seen as suspicious, evasive and not cooperative with others in the community.)

- Dependence on and yet at the same time hostility toward authority. (People in rural areas came to rely on government assistance and yet they also distrusted the government authorities they were heavily dependent upon.)

- Lack of innovativeness. (People in the countryside were seen as reluctant to adopt modernization and resistant to change; moreover, their behavior was based more in tradition and superstition than oriented in rational economic considerations.)

- Limited aspirations. (Farmers often could not imagine situations beyond what they, their fellow villagers, and their parents had known.)

- Limited view of the world. (Rural people were described as "localities" with no orientation to the world beyond their narrow group; often people were isolated with limited mobility to the outside world.)

While this school of research may indeed have identified cultural values that led to a lack of empowerment, too often the end result of such research was to lay blame on the very individuals who were, in fact, the ones who were supposed to be assisted by third world development efforts. Such research may have helped to create a stereotype of "laggards" — who refused to see the "light" of modern innovations and thus removed responsibility for failure from the innovators themselves.

Similarly, policy studies or "research" in education may at times operate on a "deficit model" where teachers may be heavily faulted for their lack of innovativeness. The history of education reform has many examples of "teacher bashing," where teachers are viewed as the main problem and they are berated for "resisting" improvements. Equally, students (particularly disadvantaged students) are sometimes seen as the problem rather than the goal. This mindset may, in fact, foster attitudes that help perpetuate rather than solve problems. Outside experts and administrators belittle teachers for being lazy and passive. More than anything else, this deficit model demonstrates the ineffectiveness of a top down change strategy.

ISSUE: EXTERNAL BARRIERS

Another school of international development researchers studied the set of external constraints that made the adoption of innovations very unlikely. Field workers in Kenya cited a variety of systemic, external, non-psychological constraints that served as bottlenecks to the adoption process:

- Lack of an effective system for delivering knowledge and skills about innovations and their advantages.

- Lack of involvement of the implementors in the development planning process.

- Lack of an effective system for delivering financial and material inputs necessary for adoption.

- Inadequate market development for realizing the economic benefits (sale/purchase) of increased output.

- Lack of infrastructure (electricity, irrigation, health facilities, communication, transportation) to facilitate communications, decision-making, and other distribution of information and materials.

These constraints sound very much like an analysis of constraints that limit the implementation of innovations in science education. Very similar barriers block attempts to increased hands-on teaching at the elementary level or the use of technology and laboratories at the high school level. A lack of financial support, infrastructure, and incentives all conspire to raise the threshold for implementing innovative practices to a level that is too high for most educators to pass over.

By identifying both internal and external constraints, researchers who studied the success of innovations began to recognize the overwhelming importance of local culture, social and economic structural inequities, and other systemic constraints. Becoming aware of the importance of constraints in both development and educational efforts is a first step away from a project-centered approach. A myriad of constraints that must be addressed suggests a systemic rather than a programmatic approach. It is not only the characteristics of the innovation — the design of the program — that are important. Rather, educators and international developers have come to realize that they have to understand and address the social, cultural, and economic conditions in which the program must operate.

To create and develop innovations requires great imagination and creativity. We thought that was enough. But in both worlds it is necessary but not sufficient to demonstrate that innovative technologies and practices offer practitioners real advantages. Now it is becoming clear that the design of efforts to help successfully integrate such innovations into daily life, to help them become part of the local culture, will demand even far greater creativity than their invention.

ISSUE: THE RICH GET RICHER

Even where innovation was successful it did not bring benefits equally to all members of society. Another important parallel can be found in the degree to which development efforts often fostered poverty and "underdevelopment" while they attempted to raise the overall economic well-being of a country. Diffusion efforts favored the already advantaged sections of the rural populations The diffusion process sponsored by the funding agencies in short produced a vicious cycle whereby the rich got richer, and the disadvantaged were largely excluded from, and ultimately hurt by, the "progress."

ISSUE: THE LOCUS OF CONTROL

Another lesson learned by international developers is that the very relationship of the developers to the developee was crucial. Top-down approaches brought in outside expertise but did not give over sufficient control or responsibility to the people who ultimately had to be empowered to use the innovations.

Robert Moses, who developed a grass-roots "Algebra Project" as a political movement that mirrored the early efforts of the civil rights movement, describes well the difference between current top-down funding strategies and a strategy based upon grass-roots community organization (*Harvard Educational Review*, Vol. 59, November 1989, pp. 27-47):

> *The community organizing approach to educational innovation differs from traditional educational interventions in several important ways. The principle of "casting down your bucket where you are" stands in marked contrast to research programs originating in universities, where scholars design interventions they hypothesize will result in outcomes they articulate in advance and that are replicable. Researchers in universities and consulting firms must have well-designed, highly articulated interventions in order to convince funding agencies that their projects have promise. Depending upon the focus of the innovation, the researcher generally targets selected neighborhoods, schools, or organizations for participation due to their demographic or similarly quantifiable characteristics. Additionally, researchers have intellectual roots in their own disciplines, and view problems through lenses that are consonant with their disciplines, rather than through the eyes of a community.*
>
> *In contrast to the university-based researcher, the organizer... gradually becomes recognized by community members as having a commitment to their well-being. The organizer immerses him- or herself in the life of the community, learning its strengths, resources, concerns and ways of conducting business. The organizer does not have a comprehensive, detailed plan for remedying a perceived problem, but takes an "evolutionary" view of his or her own role in the construction of the solution. He or she understands that*

the community's everyday concerns can be transformed into broader political questions of general import. The form they will take is not always known in advance…. It is the organizer's task to help community members air their opinions, question one another, and then build consensus — a process that usually takes a good deal of time to complete.

Moses also points out the incompatibilities of current funding strategies with a bottom-up change strategy:

Finding (financial) support can be a depleting struggle for many innovative (grass-roots) efforts…. National funding sources are hesitant to fund projects with grass-roots leadership, a community focus, a long timeframe, and a philosophy that casts educational issues in political as well as technical terms. Declining state and local budgets also threaten commitment to comprehensive, long-term reforms. But only when major political questions are addressed . . . can we discover the most appropriate ways to organize knowledge, develop curriculum, and encourage home, school and community participation.

At this time it is honest to say that neither funders, change agents, or mainstream educators know how to bring about systemic changes that result in deep and long-term change in K-12 science education. We need greater wisdom and new insights into the ways that we invest in change, the ways that we conceptualize the "projects" we fund, and even the ways that we structure our relationships with each other. We are deeply immersed in a project-centered way of thinking, and the educational community may find, like the international development community, that our rhetoric changes while our practice does not.

ELEMENTS OF NEW CHANGE STRATEGIES

At the risk of creating overly simplistic dichotomies, I offer a summary of some of the more important ways in which project-centered and systemic change strategies might differ. It is important to point out that systemic change approaches are not always superior to project approaches, or that systemic change is now the next new "answer" to the educational crisis. Rather the chart on the following pages is probably more useful as a heuristic to encourage further thinking about change strategies — thinking that will hopefully raise the clarity of all of our reform efforts.

To invest in science education reform effectively, we need to better understand the systems we are trying to change and we need more accurate ideas of how that system changes itself (or doesn't). While the conference was not successful in identifying the strategy to pursue, it did produce many good ideas about the general characteristics of effective reform efforts and investment strategies. In this section, we review and summarize the ideas that might underlie new approaches to funding science education reform efforts. We pose these ideas not as complete in themselves, but rather as challenges that must be met by funders, change agents and practitioners together.

We need to be able to redefine the relationships that currently structure most project-centered reform efforts. There is indeed a role for outside expertise and resources in assisting those working in the educational mainstream in their own development efforts. And yet the relationship between those on the outside and the inside must be both honest and forthright, allowing for local communities to draw upon outside expertise, while at the same time empowering those local communities to take responsibility for and control over their own development.

This shift in relationships is perhaps the most difficult step in moving toward a systemic approach. It involves shifts in economic and political power, as well as changes in the locus of control. Teachers and others in the schools live in a culture that indeed has some of the characteristics of the "peasant subculture." They have a mindset congruent with a long history and tradition of non-empowerment. And yet this tendency toward inertia is not a good starting point for developing successful change efforts: effective approaches will have to find ways to work from practitioner strengths and interests, not from their weaknesses or deficits.

While there is increasing belief in the funding and change agent community that a grass-roots or bottom-up approach is needed, the external change agents share a reluctance to "let go." Again, there is a strong parallel with those working in third world development. The challenge for those undertaking systemic change efforts will be to find ways to

Project-Based Change vs. Systemic Change

Project-Based Change Model	Systemic Change Model
Primary change agents are external to the system being changed	Primary change agents are key players in the system that is changing itself
Locus of control: control of the project ultimately resides in the hands of the external funders and experts	Locus of control: control of the systemic effort ultimately resides in the hands of those living and working within the system
Role of change agent: designer, leader, orchestrator, developer, outside expert	Role of change agent: facilitator, organizer, coordinator, technical assistant
Activities of (external) project members seen as central activity; grass-roots participation as a means to achieve project's goals	Grass-roots participation as central and ongoing process; project is seen as a means for enabling and supporting local processes
Deficit assumption: people need to be helped (motivation: charity)	Strength assumption: people can help themselves (motivation: shared self-interest)
Time perspective: short-term	Time perspective: long-term
Projects are temporary and develop alternatives; they operate parallel to mainstream practices	Systemic efforts are continuing and embedded in mainstream practices
Projects separate out and specialize in efforts to develop curriculum, provide professional development, or restructure schools	Systemic efforts seek integrated efforts to develop curriculum, provide professional development, and restructure schools
Projects operate within constraints of existing economics, structure and governance	Systemic efforts address key constraints – economic, regulatory, structural and cultural
Projects use schools, teachers, and students to learn better how the model should work	Systemic efforts use researchers, university experts to learn better how the system might modify its own efforts to change
Funders fund projects based on the expertise of the external change agents	Funders fund efforts based on the capacity and commitment of the system

Funders place high premium on: novelty, innovation, effectiveness	Funders place high premium on: collaboration, incremental value, capacity building, integration
Schools are seen as field or test sites; they are used to "pilot" new ideas	School sites are seen as the primary actors and beneficiaries
Underlying metaphors: engineering, product development, medicine, technology diffusion	Underlying metaphors: learning community, third world development, grass-roots community organization, agricultural extension
Projects separate the research and development phase from the dissemination effort	Research and development happen in the local system simultaneously with "dissemination" (incorporation)
An "infusion" model: projects develop and illustrate improved products and practices which (hopefully) the system will recognize, adopt, and implement	Efforts provide assistance to systems as they re-engineer (and replace) their own products and practices
Projects seek to add products and practices to current repertoire (assumption: there is a vacuum waiting to be filled)	Efforts seek to have participants decide to gradually replace products and practices with more successful ones (assumption: there is already a full plate of existing practices and products)
Assumption: reform will happen when teachers comply with recommended procedures	Assumption: reform will happen when teachers have time, resources, and permission – space to experiment and fail – and achieve authentic ownership
Project success judged by standards that focus on quality and extent of impacts, outcomes	Project success judged by standards that focus on quality of process, interactions
Evaluation, done by outside objective evaluator, yields proof of effectiveness, generalizable findings, prescription, recommendations for others	Evaluation, done by participating insider or outsider, contributes to the conversation; specific findings unlikely to be generalizable; recommendations for system participants

empower teachers (and many others involved in education) while at the same time acknowledging their realistic limits of knowledge, skills and experience. Systemic efforts will need to provide wide ranges of integrated support to help local educators see and overcome existing limits. Such arrangements only come out of honest and appropriate relationships that transcend existing structural and professional definitions.

Following the lessons learned in international development efforts, we will clearly need to identify, and then address, both internal and external constraints. Systemic efforts can address very real external constraints by helping remove regulatory constraints to innovation, and providing a range of resources that will help to build some of the systemic capacity (infrastructure) noticeably absent in our educational systems. Funders will need courage to fund capacity-building efforts, which are both necessarily vague and long-term in nature, but which are also essential to deep-rooted systemic reform efforts. Clearly, this will require changes in the reward and incentive structures of the very foundations in which the funders find themselves working.

We will also have to address internal constraints which reside in the culture of our schools, in the habits of mind of the practitioners, and the traditional thinking of the broader society. Such internal constraints are best addressed by a long-term educational effort, accompanied by a political effort to rally resources. The larger culture in which schooling takes place must be asked to shift its stereotypical views of schooling and learning. Because everyone has a personal history of schooling, it is very difficult to create alternative visions of what education might be. Public campaigns on the order of those conducted for smoking, or AIDS awareness, may well be needed. The point is that funders should not continue to fund one-dimensional programs that ignore the very real external and internal constraints that make a lie of their underlying change strategy.

Systemic efforts cannot be planned by any one person or group in advance; they must involve alliances, partnerships, and all key players and decision-makers. Ultimately, the collaborations and alliances will have to soften the distinctness of the identities of the collaborators, and boundaries will have to blur. Projects too often reinforce the professional identities of those involved, thus contributing to the fragmentation, which in some sense, is itself a core cause of the very same systemic problems being addressed.

More time needs to be spent understanding the system as it actually is now, why it is the way it is and then developing funding and reform strategies that reach into that system and change it, or help it change itself. It is not only funders and change agents who are naive and simplistic about change. In general, the practitioners working in the educational system are too close and have too little time to be insightful about the changing of their own culture and systems. The broader public is simply confused and frustrated by what they see as a huge system that appears to be both nonfunctional and unchangeable. Across the board all players need to pay attention to, learn about, and explore together the issue of change itself.

New strategies for change will not help science education unless we also arrive at a shared vision of what science education should be. To be palatable to educators, a shared vision for science education will not only have to be grounded in philosophical tenets about the nature of learning, but it will also have to be viable in the context of all of the realities of today's schools. This vision will have to illuminate clearly the nature of traditional practice, concrete examples of "idealized" practice, and key transitional practices that can be used as stepping-stones along the way. A shared vision must communicate a passion for inquiry, a love of science, and an affirmation of the importance of the reform endeavor.

To be viable in the educational community, a vision for science education must be abstract enough to allow "buy-in" on a national scale, leaving room for ongoing debate as to the exact ways that vision is to be manifested. Similarly, a vision must allow for involvement of local site ownership and further development, and not be seen as yet another "top-down" mandate.

It is also critically important that a shared vision of reform address directly the need to create learning opportunities where every student has a chance to succeed in science (and, indeed, in education more broadly). Groups traditionally underrepresented in science and in successful science education include minorities, women, students with limited English proficiency, and economically disadvantaged students. A vision of science education must directly point out the ways in which these students will have new opportunities for successful experiences.

The shared vision we create must also be comprehensible and appealing to a broader community, which, in turn, means that we also will have to transform our visions of educational change into larger political terms and social issues. Few Americans will lose sleep over the fact that their children are not engaged in hands-on science. On the other hand, if they come to understand that their children are receiving an education whose very nature fails to empower them to succeed as individuals, there is a more compelling argument for educational reform.

AMERISTROIKA AND THE ART OF SCIENCE EDUCATION

BY ART SUSSMAN

Articles in this book, funders' requests for proposals, and science education conferences all use the phrase "systemic reform" or "systemic change." Generally, we use the word systemic to indicate that all the intertwined features of the present system must change in fundamental and mutually reinforcing ways. Yet, to be honest, in our era of catchy political slogans and sound bites, most of us have developed a healthy cynicism toward the current phrases. Can systemic reform be more than this year's buzz word?

Programs described in this book demonstrate that the potential does exist for systemic reform of science education. We "simply" need to integrate many diverse elements within society in a sustained and adequately nourished program that addresses all the major aspects that constitute the present science education system. It seems so logical. Why haven't we done it already?

We just do not seem to operate that way. What we are trying to do is unprecedented, at least in America. We treat science education the same way that we handle our other problems: Wait until the crisis is so bad that it threatens to capsize our ship of state, then apply the nearest quick-fix that is available or politically popular. Look for short-term solutions that will keep the boat afloat long enough so we can rush to the other end and put a patch in the health care system or the banking system or the energy system, or whichever is the next biggest leak to hit the newspaper headlines.

How nice it would be if we had "made in America" models of successful systemic reform efforts that we could neatly apply to our situation in science education. Why don't we have good models of systemic change? Perhaps there is a very simple psychological truth — we hate change! Although we know that it is inevitable, that we grow because of it, and that, in retrospect, we are often glad that it happened, we hate change.

Many of us, especially those who are currently in rewarding or powerful positions, receive many benefits from the status quo. We have salaries, fringe benefits, defined roles, assistants, psychological reward systems, frequent-flier miles, and comfortable habits. Those of us in middle and upper management have the authority to make changes, and yet we may see ourselves as losing our advantages and power as a result of the proposed changes.

And those of us who are suffering within the system may also resist change for very simple reasons. How do we know that we are not jumping out of the frying pan into the fire? How can we work hard for positive change when our hopes have been dashed so many times before?

The challenge is truly awesome. Science educators often compare their reform task to rebuilding a 747 while we are in flight over the Antarctic. We are constrained by everything that exists. How nice if we could cancel school for a year or two while we redesign the system. But we do not have that luxury.

Can we describe all the major details of a reformed science education system? Unfortunately, that is difficult to do. Our situation is similar to that facing a writer, a sculptor or a painter who is beginning a major work, the most ambitious project of her or his career. We have a vision of our direction and destination, we know some of the tools that we will use, and know some of the major outcomes that we desire. Yet we also have learned from previous experiences that our project will take us in surprising directions. If we remain open to the fluid nature of our enterprise, the end result will be much better than if we rigidly impose too many preconceptions that prevent our efforts from evolving in fruitful directions.

With that warning in mind, we still can provide quite a few details about our enterprise. As described in other articles in this book, we know that it must be comprehensive and that it must simultaneously address all the facets of the existing system in an integrated manner. We have learned that we need to avoid simplistic solutions, that "the answer" will not come just by producing better textbooks, or from changing the tests, or issuing governmental policy statements. We need to have policy statements, curricular materials, and assessment tools that reinforce each other. Classroom lessons must foster student curiosity and enhance thinking and problem-solving skills. Assessment methods have to reward students for successfully employing skills to address new situations rather than for regurgitating isolated facts that they have temporarily memorized. And all of that is only the beginning. Among other things, we need productive parental and community involvement, a professionally empowered teaching staff, restructured classrooms, and redesigned college courses.

A good way to learn is to examine our mistakes. How lucky we are to have so many valuable lessons at our fingertips:

- Do not give money simply to start pilot programs, and then withdraw support after they have finally worked out the bugs and are beginning to provide productive results.

- Do not give money to schools from 20 different pots, each with its own complex set of rules, attached strings, and special interest groups.

- Do not announce new priorities every two years.

- Do not expect teachers to provide excellent science instruction if they have an inadequate knowledge base, overcrowded classes, few supplies, and no planning time.

- Do not expect to radically change a $250 billion system with a $3 billion investment.

Instead, make a commitment that is united rather than fragmented, sustained rather than short-lived, dependable rather than fickle, and abundant rather than deficient.

We also must become much wiser about the psychology of positive change — what enables people to make change and how people keep up their spirits in the face of seemingly overwhelming obstacles. Previously, I acknowledged one discouraging factor — we hate change.

Now it is time to acknowledge a very encouraging psychological factor — we love change. Adventures capture our imagination. Challenges spur us to undertake truly Herculean efforts.

One of the greatest incentives and rewards of participating in the science education reform movement is that our underlying philosophy is based upon belief in the human potential to live well and wisely. We build our program on the following assumptions:

- Humans are naturally curious and want to learn.

- Children learn best by exploring and developing their own understanding rather than being told what to do and being indoctrinated with the "correct" answers.

- We have enough faith in our system and in our children that we trust they will reach the best conclusions by thinking through the issues themselves and having the power to make decisions.

- If we provide the proper tools and incentives, people will design and implement radical changes that help everybody.

- All of us have skills and strengths that can be accessed and developed if we provide an appropriate environment.

- We can combine our varying skills and strengths to create a harmonious whole that is much greater than the sum of its parts.

One limiting factor is that we do not have a well-developed national consensus regarding personal growth and societal change. Consider how much time, money and knowledge that we spend on the psychology and practice of convincing people to buy a pop star's latest CD, choose between nearly identical political candidates, or use the newest "new and improved" mouthwash. Imagine how advanced we might be if we devoted 10% of that effort to the psychology and practice of motivating people to spend their time and energy truly nurturing themselves and creating a healthy society. And no, we do not want boring, moralistic messages. We are clever enough to design a missile that can find a nose cone in an underground silo 5,000 miles away. Perhaps we can be clever enough to redesign a classroom, a home heating system, or a hospital emergency room, and to enlist people's support and participation in the restructuring process.

Since we lack a consensus on how to move individuals and society in the direction of positive change, I think it would be useful to highlight a few key characteristics of the change process. We are in a Catch-22 situation. We do not believe that radical transformation can occur; therefore, real change does not happen, reinforcing our negative belief structure.

Perhaps the most significant factor is a perception that change is possible, that it is happening and that it is positive. Real change can happen only if we believe it can happen. We must act as if we are naive optimists. We must be naive optimists.

Change is autocatalytic and potentially explosive. When the process begins, it feeds on itself. No matter how well perestroika eventually succeeds, the blinding speed with which the former Soviet Union dissolved and is restructuring is a reminder that revolutionary transformations can happen today. Once people realize that things are changing, they begin to question and alter layer upon layer of practices, beliefs and assumptions. This process can lead to revisions of the most fundamental assumptions about who we are and what we do with our lives. Once the change process reaches that deep a level, almost anything can happen.

Applying these notions to systemic reform of science education, we can reach a few basic conclusions. In our beginning efforts, we should be committed visionaries who are able to convince others of wonderful possibilities even when the evidence of our senses may argue otherwise. If we as change agents and leaders lack hope, then the situation is truly hopeless. We need to highlight positive changes that are already happening. We need to find creative ways to convince the local news media to stop headlining problems and to begin highlighting solutions.

Further, the reform of science education will be most successful if it connects with other reform efforts rather than operates in isolation. Most change agents would argue that one cannot truly change precollege science education without restructuring all K-12 education. Taking the argument even further, reform of precollege science education will be most effective when it is part of an educational transformation program that includes the K-12 system, preschool, colleges and universities, adult education, and informal education centers. On an even

Status Quo	Ameristroika
Energy/Environment	
Production/Distribution by multi-national corporations	Local solutions and resources
Based on consumption of fossil fuels	Emphasizes conservation and renewable resources
Users as passive recipients	Users as educated participants
Pollution controls after problem are too big to ignore. Solving problems with technology	Pollution prevention a prime consideration in system design
Assessment – quantity of fuel consumed	Assessment – quality of life, health of ecosystems
Decisions based on corporate exigencies; individual's wealth determines services	Decisions based on health of planet and needs of local communities
Health Care	
Emphasizes high-tech, expensive solutions (e.g., triple bypass surgery)	Emphasizes low-tech inexpensive prevention (e.g., nutrition, exercise, not smoking)
Users as passive recipients	Users as educated participants
Assessment – diseases cured, years lived	Assessment – diseases prevented, quality of life
Large variations in access and quality of care	Equitable access and quality
Short-term, localized solutions to problems with the system	Systemic reform simultaneously addresses all aspects of the system
Large, centralized medical centers isolated from their local communities	Many small medical centers integrated within their local communities
Judicial System	
Emphasizes punishing criminals	Emphasizes preventing crime and providing new opportunities for offenders
Lawyers have adversarial roles	Emphasizes mediation to find win-win solutions
Large variations in access and quality of service	Equitable access and quality of service
Lawyers and judges are experts; system is complex so citizens are passive	System simplified to embody common sense; citizens are educated participants
Assessment – statistics of numbers of crimes, prisoners and prisons.	Assessment – individual and societal feeling of safety, and that the system works and is fair.

broader level, when we as citizens perceive ourselves as redefining and restructuring our health care system, our energy system, and our employment system as well as our education system, then all these efforts can powerfully reinforce, inspire and facilitate each other.

In the hope of introducing another more-than-a-buzz-word into the dialogue, I want to place systemic reform in science education within this broader context. Consider for a moment the possibility of Ameristroika — a restructuring of America. Imagine the possibilities of fundamentally restructuring major American societal institutions based upon a common vision, philosophy and methodology. Embodying Ameristroika, the changes in health care, employment, education, community social services, energy, etc. would reinforce each other. Some of the features of that vision and philosophy include:

- A community grass-roots foundation, providing power and decision making to individuals and communities.

- A focus on prevention rather than attempting to cure preventable problems.

- A systemic scope, addressing all the important factors.

- An allocation of benefits and sacrifices equitably across all sectors of society.

- A focus on emotional and spiritual fulfillment rather than purely materialistic goals.

- Conservation of limited resources and/or reliance on renewable resources to promote sustainability of the planet's ecosystems rather than their destruction.

By providing specific examples of different societal contexts, the accompanying table may also help illustrate the meaning of Ameristroika.

The reform efforts in different areas can reinforce each other in numerous ways. As already described, each helps create a climate that makes change possible. In addition to this important psychological factor, they can provide immensely important supporting benefits. Many children today come to school with a legacy of neglect, abuse, fear, alienation, violence, poor nutrition, and inadequate health care. How much easier would our job in schools be if other systemic reform efforts provided good nutrition, comprehensive health care, and communities of hope rather than despair. Children could then come to school truly prepared to continue to blossom.

Reform efforts in one area also provide valuable lessons that can be adapted to other areas. In science education we are learning how important it is to have assessment methods that reinforce the reform vision. We sabotage our efforts if we use multiple choice tests that measure only the ability to memorize and regurgitate isolated facts. These tests directly tell students, teachers and the community that it is not important to develop deep conceptual understandings of science or higher order thinking skills or science process skills. The current tests reward wrong ways to teach and learn.

We can apply our understanding of the importance of assessment in science education to other areas and discover that we use defective assessment indicators in many of our societal systems. Perhaps most gross and damaging is the Gross National Product (GNP), the indicator that is most frequently cited as a measure of the nation's economic well-being. The Exxon Valdez oil spill resulted in $1 billion of economic activity to try to reduce the environmental damage. The GNP counted this billion dollars on the positive side of the ledger! If I carpool or dry my laundry outside in the sun or turn off the television when I am not watching it, the GNP ignores these actions. On the other hand, if I drive a single-occupancy commuter car, burn fossil fuels to dry my clothes, and leave my television on 24 hours a day, the GNP goes up and tells me and society that we are doing well.

The GNP is as defective an assessment tool in measuring societal well-being as vocabulary-based multiple choice tests are in measuring students' understanding of science. We need to include factors in our societal assessment that tell us how our families are doing, what life is like in our neighborhoods, and how we feel inside. The way we assess our progress as a society needs to include the crime rate, the suicide rate, a job satisfaction survey, and health care statistics. We need to take into account our local and planetary environment — is the air clean, do we have to worry about the ozone layer or global warming? As we become more sophisticated in assessing progress in specific areas such as science education, we can apply those lessons to assessing our progress in health care, economic activity, or overall societal well-being.

Implementing the radical changes demanded by a true Ameristroika will cause severe pain and dislocation. Real change is difficult, and we will not avoid the pain by ignoring it. However, we are in much greater pain now because of the status quo. Many of us may not realize it because we are used to, and therefore unconscious of, the suffering that we have. And if we just maintain the current system, we seem destined for unprecedented disasters as we destroy our planet's life support systems.

The greatest reward of Ameristroika is that we would wake up each morning knowing that we belong to a society that helps all people reach their highest potential, and that aims to live sustainably within the resources of our planet. People may be willing to make sacrifices to realize that vision. If they perceive that everyone shares equitably in the pain and in the gains, people will surprise you with the sacrifices that they make.

Science education partnerships can fit well within this Ameristroika framework. However, to be honest, even the best of them today do not live up to their potential. They fight for survival and success against tremendous odds. One of their biggest unrecognized limitations is that they operate too much from the perspective of changing the schools, of a flow of services from a science-rich institution to needy schools and teachers. For science education partnerships to be truly effective and promote the radical changes that we envision, the science-rich institution (college, university, informal science education center, or industry) must be equally willing to radically change.

This lesson is easiest to see in partnerships between higher education institutions and K-12 schools. Teachers teach the way they were taught. Colleges must change their undergraduate science courses, especially those for prospective teachers and for the non-science major. To encourage excellence in teaching, university faculty must be rewarded at least as much for public service and undergraduate education as they are for academic research. When universities confront their own transformation needs, they will be in a much better position to learn from and contribute to systemic reform of precollege science education.

Over two centuries ago, a group of white men initiated a systemic societal reform that came to be known as the United States of America. Today we are a coalition of men and women of diverse colors and nationalities who are challenged to create a better world for our children and for each other. As I write this essay, my laptop computer is a reminder that we have previously unimagined tools at our disposal.

I have attempted to show that science education partnerships can contribute to and benefit from a much larger vision of societal reform. Whether we view them from this larger perspective of societal change or simply for the good that they accomplish locally, science education partnerships demand our best efforts and provide us with many rewards. In the face of seemingly overwhelming odds, it may be difficult to feel good about the limited contributions that we can make or to maintain our vision of systemic reform. More than 100 years ago, Thoreau drew upon his own experiences to reach conclusions that may still inspire us today:

"I learned this, at least, by my experiment: that if one advances confidently in the direction of his dreams, and endeavors to live the life which he has imagined, he will meet with a success unexpected in common hours. He will put some things behind, will pass an invisible boundary; new, universal, and more liberal laws will begin to establish themselves within and around him; or the old laws be expanded, and be interpreted in his favor in a more liberal sense, and he will live with the license of a higher order of beings.... If you have built castles in the air, your work need not be lost; that is where they should be. Now put the foundations under them."

ORDER FORM

SCIENCE EDUCATION PARTNERSHIPS

MANUAL FOR SCIENTISTS AND K-12 TEACHERS

Edited by Art Sussman, Ph.D.
Preface by Bruce M. Alberts, Ph.D.
President, National Academy of Sciences

Published by University of California, San Francisco

Science Education Partnerships: Manual for Scientists and K-12 Teachers features 35 articles that describe a multitude of collaborative programs that are playing a vital role in the improvement of precollege science education.

Please send me *Science Education Partnerships: Manual for Scientists and K-12 Teachers*

Quantity   Amount

_____Paperback at $14.95 each   _____

_____Hardcover at $24.95 each   _____

Add $4.00 for shipping and handling (no matter   _____
how many books). Foreign orders, add an additional $5.00

CA/Canada Sales tax _____

Total _____

Please check one of the following:

☐ I am enclosing a check payable to Science Press

☐ I am enclosing purchase order number _____

Name _____

Address _____

City _____ State _____ Zip _____

SEND ALL ORDERS TO:
Science Press, P.O. Box 31188, San Francisco, CA 94131
Purchase orders may be FAX'ed to (415) 476-9926
Attention: Science Partnership Book